In Memoriam
Einstein

*The Einstein Centennial Symposium
at the Institute for Advanced Study
March 14–18, 1979*

Roy Lisker

A Ferment Press Book

© 1979, 1987, 2005, 2015 by Roy Lisker
Book design © 2015 by Sagging Meniscus Press

ALL RIGHTS RESERVED.

Set in Adobe Garamond with LaTeX.
Cover design by Anne Marie Hantho.

ISBN: 978-0-9861445-9-2 (paperback)
ISBN: 978-1-944697-00-6 (ebook)

Sagging Meniscus Press
web: http://www.saggingmeniscus.com/
email: info@saggingmeniscus.com

Average men find communication easy, whereas intellectuals seem to agree to erect barriers against each other which they are unable to remove. This shows that civilization contains an inhuman element.

—Marie Curie

Contents

The Event Horizon · 1

History and Geometry	1
Insights into the Author's Past	5
The Einstein Revolution	7
Arrival in Princeton	14
I am Arrested	15
I Become a Journalist	16
Finding Room and Board	19

The Symposium · 21

The Opening Ceremonies	21
Historical Illiteracy	22
Some Strangeness in the Proportion	25
Einstein, Where Were You?	29

Blueprint for a Symposium	31
The Einstein Age: 1900–1926	34
Max Plank Defenestrated	37
Albert Einstein Defenestrated	38
Dinner with Subramanyan Chandrasekhar	40
Lunch With the Bureaucrats	41
Experimental Relativity: Irwin Shapiro	42
The Cocktail Party	44
Social Responsibility at the IAS	46

Cosmology · 47

Sciama's Visit to Cygnus X-1	47
The Battle of Princeton	49
Upstairs, Downstairs at the IAS	52
Black Hole Phrenology	54
Eternal Life and the Baryon Number	56
Fields and Particles	57
Isaac Rabi Resuscitates the Media	59
The Cosmological Principle	61
Summary of Cosmology	65

Finale · 67

Einstein on the Lunch Menu   67
Quantum Gravity; Supergravity   69
Relativity's Future   75
I Become a Filmmaker   76

Homecoming · 79

Bibliography · 83

IN MEMORIAM EINSTEIN

The Event Horizon

History and Geometry

The study of history might benefit through adopting the viewpoint commonly held in General Relativity. The hypothetical historians interested in this approach should have the requisite training in mathematics and physics. That this programme be more than a fatuous exercise it needs to be applied both to the investigation of the historical forces that shape collective behavior, and (analogous to General Relativity's treatment of matter and space) to the reciprocity between personal ambitions and national destinies. The principle of equivalence asserts that what we measure as a gravitational force between massive objects can better be understood as a distance measurement in an immutable 4-dimensional space-time geometry. Although this geometry is described by Einstein's field equations, there is no consensus about the shape of its universe. Many candidates exist, the so-called models of Einstein's gravitational equations: the "special" approximation of Minkowski Space; de Sitter Space; anti-de Sitter Space; Reissner-Nordstrom Space; Robertson-Walker Space; the Schwarzschild Space; Taub-NUT Space; Kerr Space; Gödel Space. Time is "static" in Gödel space, which is what one might expect from the Master of Undecidability.

I. The Event Horizon

The "plumed serpent" Quetzalcoatl, found this out to his cost:

> The Mexican snake Quetzalcoatl
> Got stuck in an open Klein bottle
> My word, am I dotty?
> I've gone through my body!
> And feel like a theorem by Gödel!

In these models (of which the Robertson-Walker and the Schwarzschild are the only ones used in most applications), massive objects no more "attract" each other, than does the fact that one side of a triangle has to be shorter than the sum of the other two indicate that there is an "attraction" between the triangle's vertices.

From this Parmenidian perspective, the illusion of activity cedes place to the permanence of inviolable principles. In a relativistic theory, things appearing to move are in reality gliding down a slope of least resistance, the so-called geodesic in spacetime. This isn't motion to relativists: such paths are called flat.

Applying this viewpoint to a theory of history, isolated events (individual, collective, national, even global) would be described in terms of distances of a static history metric. World-shaking events, mass movements, revolutions, even fads and fashions, could be examined as curious geometric shapes in history space. Phenomena as disparate as the emperor Constantine's conversion to Christianity and the building of the giant statues on Easter Island could be "deduced" from a small set of postulates similar to those of Euclidean Geometry:

- Triangles are congruent if two sides and included angle are equal;
- The area of a circle is 3.14159… times the radius;

History and Geometry

- The square of the hypotenuse of a right triangle equals the sum of the squares of the other two sides....

This way of looking at the world has entered into my own speculations concerning the roots of my decision to hitch-hike to the Einstein Centennial Symposium, the festive conference in honor of the hundredth anniversary of Einstein's birth, between March 14th and 18th at the Institute for Advanced Study in Princeton. How could I place my personal decision into the wider historical context? Were my decision and subsequent actions somehow inevitable, given the circumstances of my life and the political realities of the 20th century? Although this decision was quickly made after browsing over an article in the New York Times on the opening day of the Symposium, subsequent events demonstrated that it arose from a profound source.

Could I have done otherwise? Was the eventual publication of this report in a prestigious French magazine as inevitable as the Symposium itself, as my attendance there, as, ultimately, Einstein's creation of Relativity itself? These reflections are the beneficiaries of hindsight, in which all things fall into place. It is not surprising that I picture all the occurrences surrounding the event as a geometry diagram in historic space-time.

Yet, on the morning of Sunday, March 14th, 1979, when I came down for breakfast in the dining-room of the commune in the Hudson Valley where I'd washed up in the late 70s, I had no idea that this date was the 100th anniversary of Einstein's birth. My previous sojourns in Princeton were decades in the past: there was no-one, to my knowledge, either on the campus or in the town to whom I could turn for help.

I. The Event Horizon

After reading through the article explaining that a host of famous and prominent people would be gathering at the IAS for the Symposium, I called up a friend, the physicist Peter Skiff, teacher at Bard College a few miles down the road, to ask for his opinion. He tried to discourage me from going. He liked me, he said (I'd never thought otherwise) and didn't want my feelings to be hurt. Even if I somehow managed to gain admission (which he believed most unlikely) I would be snubbed by all. To his mind the whole extravaganza was a media event, an exercise in self-aggrandizement by an elite, self-infatuated club of winners, runner-ups and would-be winners of Nobel Prizes. Their only reason for coming together was to hype each other's reputations. They would have little time to waste on outsiders. Dr. Skiff was wrong, though one could understand and empathize with his bitterness. What I discovered was a world not bereft of humanity, though with its own peculiarities, of which you are about to learn.

I had no money. This is not a metaphor: the commune frowned on this widely dispersed technology. A sympathetic friend passed me $5. At around 10 AM, carrying briefcase and backpack, I walked out to Route 9E and put out my thumb. It was as simple as that. Well, no, I also held a piece of cardboard on which I'd written, in block letters with a felt pen:

> ## PRINCETON, NJ
> ### Institute for Advanced Study

There was a light rain that persisted at more or less the same intensity throughout the 5 hours I was on the road. It let up soon after my arrival in Princeton. I'd been out of touch with the world scientific community for a good 15 years. The normal questions which would trouble anyone also occurred to me; I just didn't let them get to me: Where would I stay? Wouldn't I have to eat? Gaining admission at the Symposium wouldn't be easy. What would happen if I were discovered? What if the journey took all day and was unable to get beyond New York City before nightfall?

Insights into the Author's Past

Under normal circumstances any one of these stumbling blocks would have prompted me to cancel the journey. But these were not normal circumstances. I wasn't actually making a journey, rather the journey was making itself with me as agent, a spacetime geodesic like the motion of the Earth in orbit around the sun. Dominating my consciousness was the spiritual state that characterizes all of the hardship-ridden pilgrimages of the devout, the fanatical, and the benighted throughout human history. One finds nothing far-fetched in the tale of a pious Moslem who, for no apparent reason, abandons all concern for his life, family or business and starts out on his long delayed pilgrimage to Mecca!

As with many scientists, my veneration for Einstein began at an early age. But despite my growing up in Philadelphia, a city less than 40 miles from Princeton, it never occurred to me to go out there and try to meet him. One doesn't want to meet one's idols, either from fear of misbehaving or fear of tarnishing the mystique. Once however I did come very close to meeting him.

The court is out, and may always be so, as to whether I really was the whiz-kid that the University of Pennsylvania mathematics department saw in me. The important point

I. The Event Horizon

is I was thought to be one, because of which I was skipped, at age 15, short a year and a half of graduating high school, and enrolled into a graduate program in mathematics at Penn in the Fall Term of 1954.

The question of whether Lisker would ever do first-rate research became moot when, in my sophomore year, he developed a revulsion towards everything having to do with math that lasted 25 years. Curiously, it was through my attendance at the ECS that my interest in math and physics was rekindled, by which time however I was well past the age at which 90% of all mathematicians produce 90% of all major results. That doesn't upset me overly much. Mankind had little use for mathematics between the 3rd and 13th centuries AD, which seems to indicate that other things may also be important.

As a way of hedging its bets, the Penn mathematics department sent me out to Princeton to be interviewed by Bruria Kaufman. She was Einstein's mathematics secretary at the time, wife of the well-known linguist Zellig Harris. She interviewed me, decided that I knew enough by age 15 to be in Penn's graduate program, and probably fairly smart as well (though there is no guaranteed way of knowing that). So she sent me back to Penn with an "A" on my forehead. The source of all my future woes!

There is no doubt that I could have arranged a meeting with Albert Einstein through her. Age 15 is not a time for encountering living legends, not for most of us. Hero worship is a universal vice, and I can report that I did experience a certain amount of awkwardness sitting down at a table for lunch in the company of some of the superstars at the ECS. This handicap wore off after a few days, after which I was able to chat with them with the same nonchalance I customarily brought to the motley crew of truck drivers, clerks, students, cops and prostitutes in the local restaurants which I frequented at the time.

My hitch-hiking took me down the New York State Thruway and the New Jersey's Garden State Parkway for all of that grey March afternoon. I recall having been picked up by 5 drivers. Half an hour was the usual wait between lifts, though I was stuck for over an hour at a location on the Garden State Parkway. Unrealistically I'd imagined that I would be in Princeton by 1; it was close to 4 when my final driver dropped me off in front of the Princeton Inn.

Impervious to the rains, I was fired up by the thought that the handiwork on my cardboard sign flashed around the world much like the marathon torch that precedes the opening of the Olympics. Was it not self-evident that all who saw me on the highways realized that we were living out the final day of the Einstein century, their acknowledgements shooting like telegraph messages through the collective consciousness of the human race?

Eventually even I was forced to recognize that few persons took notice of the date March 14th, 1979. Can one imagine it, there are probably millions of souls out there to whom the very name "Einstein" means little or nothing!

The Einstein Revolution

Which is beside the point. The revolutions in physics (and all the sciences) since the turn of the 20th century have profoundly affected the lives of everyone on earth, the continuation of a process that has been going on for over 300 years. No educated person need be reminded of the powerful upheavals in politics, ideology, technology, economics and life expectancy associated with the names of Copernicus, Galileo, Newton, Lavoisier, Pasteur, Darwin, Marx, Planck, Einstein. With each major increase in mankind's

I. The Event Horizon

power over nature there has been a corresponding belittling of Mankind's stature relative to the cosmos.

In some ways the revolutions precipitated by Relativity and the Quantum Theory and Relativity are unique. Previous to them no-one thought to question the objectivity of the observer relative to the observed: neither Copernicus' displacement of the center of the solar system to the Sun, nor Newton's magisterial explication of the Natural Order, nor Maxwell's Fields, nor the ultimate humiliation (to some) of Darwin's revelation of biological evolution.

However great the offense to anthropocentrism, every scientific theory up until 1900 continued to assume the existence of Observation as being independent from the Observer. Newton defined Force as a composite of space, time and matter. Dalton explained chemical properties by appealing to small, very hard atoms that would ultimately, whether directly or indirectly, become visible. Since waves have to propagate through something, James Clerk Maxwell and those before and after him up to Einstein assumed that there had to be a medium for light, which was christened the ether. Light itself had to be either a wave or a particle; it could not be both. Newton's particle paradigm had a brief life before being replaced by Huyghens' waves, which explained more.

The middle of the 19th century saw the introduction of a new mode of thinking. There were things that could never be known pragmatically; then things that could never be known theoretically. Carnot showed that the work involved in producing heat could not, even in theory, be recovered. Gauss showed that most angles cannot be trisected by ruler and compass. Galois showed that the roots of the general algebraic polynomial of the 5th degree and higher could not be written down as an expression in fractions and radicals. Cantor confounded our commonsense notions of infinity, replacing them by hierarchies without end. Darwin showed that the human race is merely an effect in a long chain of causation through the animal kingdom going back billions of years. Then, at the turn of the century, Max Planck and Werner Heisenberg proclaimed that whether we see a wave or a particle in the sub-atomic universe depends on the way we look at it.

Among all those who have led the way towards the cognitive paradigm that has since become customary in all the sciences, Einstein was unquestionably the leader. Einstein, and those who have followed in his footsteps, began describing the universe in terms of entities that "exist" only in the mathematics!

The time coordinate of Special Relativity is represented by a purely imaginary parameter, ict, where $i = \sqrt{-1}$, the square root of minus 1. Length and time are not independent; the rate at which the hands of a clock move (as seen by someone at rest) is affected by the motion of the clock itself through space. Borrowing a page from René Descartes (who rejected the possibility of action at a distance), only collisions are simultaneous; there is no before and after. All velocities less than that of light are relative, but light moves at an absolute speed, independent of reference frame. To draw a consistent picture that includes these seemingly contradictory features, one must develop a new geometry, one with lines and points that don't "exist," properly speaking, in our everyday world, but which tell us what is happening in it. Even rigid bodies, a mainstay of the old physics, no longer exist, properly speaking, in Special Relativity. (They are returned in a sense in General Relativity, their rigidity considerably mangled!)

In the General theory of Relativity, developed during WWI, Einstein replaced both matter and energy by a Matter-Energy Tensor. Tensors are purely mathematical objects, generalized vectors compounded with so-called

I. The Event Horizon

dual vectors or co-vectors. *All of these things derive out of Nature and lead back to Nature, but they do not exist in Nature.*

The propagation of light-waves through empty space is counter-intuitive; yet it is much easier to describe the behavior of light in the language of Special Relativity, than it is to describe the propagation of ordinary sound waves through water or air. One doesn't have to worry about the quantities that specify the kind of water or air.

This predilection for replacing virtually everything we see and touch by purely mathematical abstractions also became the programme of Quantum Theory, leading to a host of arcane subjects in which Quantum Theory, Special Relativity and General Relativity are blended as alloys: Gauge Theory, Quantum Electrodynamics, Quantum Chromodynamics, Quantum Field Theory, Axiomatic Quantum Theory, Quantum Gravity….

The Schrödinger ψ-function measures the propagation, through time and space, of a purely mathematical entity, like a catalyst that enters into a chemical reaction to emerge unscathed. Yet it contains "all of the information one can possibly extract" about all the observables of the real world: momenta, energies, times, positions, spins. In recent theories of Quantum Gravity, even Space-Time itself becomes a "foam of uncertainty" in the neighborhood of a Black Hole, itself an object that can only be described through the abstractions of Differential Geometry.

This is the *real world* according to modern physics. Virtually all of it is unpicturable. It was Einstein who taught us this way of looking at reality. A return to former simplicity is unthinkable. After Einstein, physics, mathematics and philosophy must forever be full partners in a triple marriage. Even many physicists are slow to grasp this, refusing to acknowledge that it is their obligation to take epistemology seriously.

The Einstein Revolution

One is speaking of a revolution in thought more extensive than those launched by Darwin, Marx or Freud. Darwin transformed our vision of the past; Relativity transforms our vision of time itself. The entities of psychoanalysis are unpicturable because their formulation entails the worst kind of pseudoscience. One cannot hope for better science than the theories of relativity. Even Karl Popper, in *The Logic of Scientific Discovery*, expresses admiration for its stubborn resistance to "falsification." Marxism, radical in its analysis of Capitalism, utilizes a conservative mechanism for historical causation. Relativity and Quantum Mechanics have reshaped the way, which goes back to Aristotle, we think about cause and effect.

Standing in ironic contrast to the challenges of this newly revealed universe were the attitudes of extreme conservatism of many of the scientists at the ECS. It is something of a paradox that I will also be describing them as bold and audacious men (with a dismal smattering of women), true revolutionaries accustomed to perpetrating the most outrageous feats of intellectual vision against public incomprehension, and even ridicule.

One can overstate the case in both directions: it is as wrongheaded to credit the revolution in physics to the miraculous intervention of a mythical demigod, as it is to debunk his contribution on the grounds that others had already done the work, or would have done so. Observe that all sides of this issue were debated to the full measure of their irrelevance in the lectures and discussion periods at the ECS. Big Bangs may have their place in Cosmology but little is to be gained from locating the origins of modern science as the Einsteinian Big Bang. Thomas Kuhn in *The Structure of Scientific Revolutions* has taught us that revolutions in thought, as in politics, only come when they are long overdue. It is the rapidity with which they overwhelm the Old Order that furnishes the illusion of spon-

I. The Event Horizon

taneity. Seen in retrospect, the proportion of cause to effect in their makeup is about that of the "moi" to the "deluge" that follows "aprés."

Herein lies the real value in the numerous Einstein celebrations now taking place around the world. A re-assessment of the past 80 years of scientific history is long overdue. Here is a brief sketch of what, to me, ought to be the principal constituents of this re-assessment:

It is commonly recognized that 20th century physics is the harvest of seeds planted in the 19th. All of the mathematical tools employed in Relativity (both Special and General) and Quantum Theory were ready at hand when Einstein, Lorentz, Heisenberg, Schrödinger and others needed to make use of them.

It is a permanent feature of applied mathematics that one never knows in advance which branches of mathematics will turn out to be relevant to the problem at hand. An example: Heisenberg's Matrix Mechanics introduced into physics a certain kind of generalized geometrical entity, matrices, invented by the British mathematician Cayley. That they originated in England is itself worthy of note: throughout the 18th and 19th centuries English mathematicians were primarily geometers. The growth of Analysis took place on the Continent.

This absurd dichotomy did not arise from any intrinsic difference in national character. Rather it smouldered on as a legacy of the bitter disputes between Newton and Leibniz over their priority in the invention of calculus. Recall the infamous "commission" sent to Germany to investigate Leibniz' claims, a fraud devised by Newton himself, yet whose report was published in the Proceedings of the Royal Society! In keeping with a tradition dating back to Alexandria in the 3rd century BC, Newton's methodology in the *Principia* is geometric. Leibniz, benefiting from the advances in notation starting with Ramon Llull, al-Khwarizmi, Cardan, Leonardo of Pisa and Viète, used a modern algebraic notation more suitable to the analytic geometry of Descartes.

Many historians consider that British mathematics was thereby impoverished from the 18th century to the beginning of the 20th. Perhaps: yet there were valuable insights into the relationship of the mind to the world in the study of geometry that could be passed over in the analytic approach: Cayley's matrices; Hamiltonian mechanics; quaternions; vector analysis; Maxwell's concept of the field; much of Abstract Algebra, which can equally be labeled Abstract Geometry. Everything in the above list was picked up in the new physics of Heisenberg, Schrödinger and Einstein.

It is not enough to cite the "great names" alongside the lists of their achievements. One must also study the scientific politics, and the national politics of the age in which they worked. In particular one must examine the effects of nationalism (Newton/Leibniz/Descartes in the first half of the 18th century) and religion (Galileo) on the content of science. These exceedingly petty priority squabbles often conceal deeper issues.

It was my hope that the Einstein Centennial Year would inaugurate a thorough retrospective of the previous century in science. We ought to know more about where we've come from, because in many respects we still don't know where we've arrived. We continue to lug about old baggage, prejudices and fixed ideas along with the latest novelties and discoveries. More than one fundamental insight has been cast by the wayside in the mad rush to progress.

It was at the Einstein Centennial Symposium that I encountered the new ideas in theoretical physics. Surprisingly, some of them have a delightful quality of *déjà vu*, thinly veiled recyclings of earlier theories, formerly discarded, resurrected in a more contemporary language. One thinks of string theory, picking up on the old "knot and vortex" paradigms of Tait and Kelvin in the 19th century.

At the same time well-established notions still have problems associated with them: quanta, fields, space-time curvature, Black Holes, non-locality…. We know how to talk about them, but don't fully understand them, even today. A historical retrospective is required to put this unruly house in order, although we may never be able to say much more about these objects other than that they are there.

Arrival in Princeton

A measure of the gulf which separates the world I entered when I walked through the doors of the IAS that afternoon around 4, and the world that brought me to them, can be gauged from my interaction with the driver of my final lift. He was a friendly elderly handyman. He'd picked me up on the outskirts of Princeton and drove me to the Princeton Inn, less than half a mile from the Institute. When I told him where I was headed he asked me to try to explain the theory of relativity to him in language that he could understand.

After a few earnest attempts it was clear that the effort was useless. No-one without at least some familiarity with the calculus can be expected to form a mental image of the relativistic universe from a 5-minute exposition. As I stepped out of the car he grasped my right hand in his, beamed his eyes into mine, and said: "I don't know anything about Relativity, young feller, but I do know this: You will never find true happiness until the day when you accept the Lord Jesus Christ as Son of God and your personal Saviour! He's changed my life and he can change yours. So, be a good scientist, and God bless you!"

I am Arrested

Rather than circling the Princeton Inn to pick up the road to the IAS, I strode through the front door, traversed the lobby to another door at the back, and stepped onto the terrace. There I climbed over a low stone wall, jumped down onto a field, and walked through a grassy meadow towards a parking lot. This was a mistake in a place like Princeton; it is the town, after all, where neighbors gossiped whenever Albert Einstein had the nerve to promenade the streets eating an ice-cream cone. Within minutes a plain-clothes cop was onto me. He signaled me to a halt and corralled me into a niche on one side of the building.

Where did I come from? What was I doing in that neighborhood? Why did I rush through the Inn and jump over the wall? Why was I running across the field? What was in my briefcase?

The cop continued to interrogate me in this most unfriendly manner. As he did so he opened my briefcase and pulled out the photocopies of 3 papers in theoretical physics I'd intended to pass along to delegates at the conference. The ultimate horror: *the man was a professor!* You don't treat professors like unwanted strangers. Heavens, you don't even treat them like students! Red-faced and stammering, apoplectic with apology, he waved me off with competent instructions about getting through the parking lot and turning onto a side road to get to the Institute.

It was getting late and I was beginning to worry: fast approaching 4 PM and 3 hours after my estimated arrival time. In fact I could not have arrived at a better time, but

I. The Event Horizon

I had no way of knowing that. The security guard that I encountered in the lobby of Fuld Hall directed me down the hall:

"Right this way sir. It's in the little auditorium. You can't miss it."

Not knowing what "it" referred to, I bustled through another the door and across the catwalk into a tinker-toy installation, the Henry Chauncey Conference Annex, the construction of glass, steel and grey stone where I would be spending the greater part of my waking time over the next 5 days. The small auditorium is in the office building adjoining the conference center. Opening a door, I stepped into the Press Conference with the casual air of a seasoned journalist. I settled into a soft seat at the back and, with a display of quiet efficiency, lifted pen and stenographic pad from my briefcase. Journalist I had become, and journalist I would remain for the duration!

I Become a Journalist

Harry Woolf, historian of science and current director of the IAS, stood behind the podium at the front of the auditorium. He managed the press corps like a professional. His answers were predictably trite but then again so were the questions. Woolf did little more than repeat clichés taken directly from the press folders given out later that afternoon, stock phrases destined to be repeated a dozen times that evening in the speeches of the dignitaries assembled for the opening ceremonies.

The press conference did contain some moments of comic relief. One ancient hack seated near the front asked: "What is your opinion of Eisenhower's place in world history?" The gentleman was dozing of course. In comparison

I Become a Journalist

with his colleagues, the question at last showed imagination. Rather than sweating out an extemporaneous reply, Woolf gently let him know that he'd committed a slip of the tongue—("You meant Einstein, sir, didn't you?")—of the sort that would have delighted Freud. Laughter rippled around the auditorium like the transit of Venus across the face of the sun. *(See Bibliography.)*

So as to lend credence to my new profession, it behooved me to ask a question of my own. In any case I suffer from a nagging compulsion to ask at least one question at most lectures. The effect of this, particularly in the academic world, has been to considerably enhance my reputation for daring.

I asked Dr. Woolf if there were any plans afoot for making 1979 a target year for advances in theoretical physics. Woolf, not aware that he'd been caught off-guard, started to rehash the verbiage in a press release about the relevance of Relativity to modern science. I had to stop him to rephrase the question: were there plans afoot to make 1979 a year of crash programs for attacking unsolved problems in General Relativity?

He didn't know the answer: perhaps this would be addressed during the conference. As far as he knew there was no such program at the Institute. (As I was to discover this was an excellent question: the answer was a resounding Yes!)

The press conference over, we all shuffled upstairs to register with the PR director for the IAS, Bill Wing. He was sitting behind a card table handing out press kits and writing up identity badges. As the line advanced I groped about desperately for names of magazines, newspapers or press agencies with whom I might claim some plausible connection. The media organ would have to be prestigious enough to satisfy the officials of the IAS, yet sufficiently remote to make them reluctant to inquire more closely.

I. The Event Horizon

I recalled the names of some small magazines in England, France and Ireland for which I'd written book reviews in the 60s. These were unlikely to impress anyone. It was also easy to dismiss the tiny Peace Movement screeds for which I'd scribbled polemics in the same period. Then there was the engineering publication I'd worked on over the previous summer in Ann Arbor, Michigan: *Appro-Tech*, the official journal of the *American Association for Appropriate Technology in Developing Countries* (Triple ATDC). Two articles of mine were published in the two issues produced before it went defunct.

Suddenly, seized with that quality of spontaneous illumination that Einstein claimed to have experienced when he grasped the nature of the photoelectric effect, I knew what I was going to say. Divested of all fear, I stepped up to the table and waited for Bill Wing to ask me for the name of my employer:

"*Les Temps Modernes*, Paris, France."

This was not a lie. Rather it was a legitimate extension of the truth in much the same way that, for example, concepts in physics are extended across new fields of scientific inquiry. *Les Temps Modernes* is the magazine founded, and edited at the time, by Jean-Paul Sartre and Simone de Beauvoir. During my years in France, between 1967 and 1972, it published 3 large articles of mine. Therefore the chances that it would accept another article from me on a subject of this importance were excellent. *Les Temps Modernes* doesn't copyright its articles, and the English language version could be submitted elsewhere. ("In Memoriam Einstein" was published by *Les Temps Modernes* in January 1980.)

The strategy worked to perfection. With an undissimulated pleasure that stopped short of gloating, I relished every moment Mr. Wing spent in spelling out, letter for letter, what was for him a difficult name in a foreign language. My standing as a journalist was now officially acknowledged by an unimpeachable document. Not only was I entitled to attend all the lectures but, as I learned the next day to my boundless delectation, to three 5-star meals every day and admission to the cocktail parties.

At the Symposium the name of *Les Temps Modernes* worked the same kind of magic that titles and affiliations do at any academic event. Scholars in the know had only to glance at my ID badge to conjure up the twin mantles of Sartre and de Beauvoir, archangels hovering at my sides in defense of my right to be at their Symposium. Because of this badge, few if any hackles were raised when I presumed to participate in the discussion sessions, normally a species of blasphemy for any mere journalist. It has been observed before that powerful credentials at an academic conference may suffice to transform a dumb question into a deep one, even as the possession of invidious credentials can transform an intelligent question into an embarrassing gaffe.

Finding Room and Board

Registration was over. It was almost 5 PM, fast approaching twilight for mid-March, and I had lots of things to worry about. Opening Ceremonies were scheduled for 9 PM. Over the next 4 hours I had to find room and board, if not for the entire conference then at least for that night.

As stated before I didn't know a soul in Princeton. I did know my way around large American universities. Leaving the IAS I hurriedly strode the two miles to downtown Princeton, the commercial district on Nassau Street just opposite the main gates of the university. In a short time I found a cheap restaurant named Buxton's. Stepping

I. The Event Horizon

through a room dingy with dense clouds of grease I seated myself at a table. A waitress arrived. She poured me a cup of coffee right away. 15 minutes later she returned carrying one of the most dreadful hamburgers ever ordained to be set before a member of the human species. Munching over my first meal since breakfast, I opened the blue folder I'd been given and started reading the press materials.

It was not 5 minutes before a pair of Princeton students, a young man and woman, stopped by my table and asked if I were involved with the Einstein Centennial Symposium. This was the magical moment I'd anticipated: my anxieties vanished in a flash. David Hastings, the young man, indicated the way to Murray Dodge Hall, the student activities building where all the religious and political groups are located.

Over the next hour I made 4 visits to Murray Dodge. Finally, around 6:30, I encountered a pair of foreign students. They called up friends, Lauren, a mathematics TA, and his companion Janet, both grad students from England. Lauren and Janet put me up for 4 nights (Sunday to Wednesday) of the conference. They also provided meals until the discovery was made that the Institute's cafeterias were open to me. My hosts turned my "pilgrimage to the source" into a reality. Through them it became a true homecoming. This article could never have been written were it not for new friends, they among others, who assisted me at every stage of the journey.

The Symposium

The Opening Ceremonies

At any international science conference politics operates at two levels, overt and covert. Generally speaking the overt politics are pursued by the governments that sponsor them or the financial interests which underwrite them. The overt politics of the Einstein Centennial Symposium were restricted to the Opening and Closing Ceremonies that Sunday night and on the following Thursday night. Neither I nor most of the participants stayed to attend the latter, a speech by the governor of New Jersey and the a message sent by President Jimmy Carter and read by Frank Press, Carter's advisor on science and technology.

The *covert* politics can be found in the humorless (therefore readily parodied) jockeying for status, prestige and jobs within the academic community. It thrives through the positing and buttressing of essentially arbitrary prejudices against unfashionable schools, paradigms, theories or individuals; and through subtle and not-so-subtle (even crude) denigration, exclusion, put-downs and snubs.

Most of the delegates at a science conference are only interested in the covert politics. That they can be extraordinarily deaf to the overt politics which, time and again, manipulates these prestigious gatherings to its own ends is one of the major tragedies of the 20th century. During the 70s I never ceased to be astonished to read about all those *eminent and distinguished men of science* who lent their authority to the "International Science Conferences" organized by the Unification Church of the "Reverend" Sun-Myung Moon. Were the Ayatollah Khomeini or Augusto Pinochet to offer them hospitality, one would witness the sorry sight of these esteemed sages flying to Iran or Chile, bestowing their academic renown and the universal respect mankind bestows upon them onto these notorious regimes. There are other ways to respond to such offers: one recalls Arturo Toscanini's reply to Hitler's offer that he conduct the Bayreuth Festival: *One does not co-operate with evil.*

Not that scientists are either more or less opportunistic than the rest of us. It is amazing, all the same, to see how far so many of them will take refuge in the old chestnut about the "autonomous and eternal virtue inherent in pure scientific research." And to speak truly, all the time they are at such conferences, they will be conducting their perennial in-fighting, the often nasty covert politics within their profession, fatuously convinced that this is the only kind of politics worth considering.

Historical Illiteracy

The Opening Ceremonies at the ECS were marred by a phenomenon which I've labeled historical illiteracy: the inability to read the historical significance of the event in which one is a participant. Leni Riefenstahl is its paradigm. It isn't quite the same thing, I suppose, as cynicism or indifference, in the same way that we don't consider the inability to read books a form of cynicism or indifference.

Historical Illiteracy

The ECS from beginning to end was riddled with historical illiteracy. That the official press corps delegated to cover the conference was, by and large, hopelessly inept, can be considered another one of its manifestations. It was also present in the triage performed by the 8-member Planning Committee (Harry Woolf, Freeman Dyson, Herman Feshbach, Marvin Goldberger, Gerald Holton, Martin Klein, Abraham Pais and John Wheeler) which determined a participation structured by biases one would have considered unthinkable in the late 70s.

Yet it was at the Opening Ceremonies that historical illiteracy was most blatant. One heard it in almost every phrase of every speeches. One saw it in the audience who responded, like brainwashed zealots, with mindless applause. It was actually a sign of health that the Opening Ceremonies were not only overtly political, they were entirely political. They didn't have a thing to do with Einstein, though they had a great deal to do with political relations between West Germany and the United States. One shudders at the thought that Albert Einstein himself might somehow enter into this centennial celebration in his honor, only to discover that his archetypal persona, that of the homeless intellectual fleeing the cruelest persecution in history, was being used as a springboard for sleazy political gestures between powerful capitalist nations and their scientific establishments. Apart from the speeches, the Opening Ceremonies of the ECS consisted of three events and one notable omission:

(1) The presentation to the Institute for Advanced Study of a sculpture by Jacques Lipschitz, "Arrival." It was conceived and produced in 1933 as a poem of thanksgiving for having escaped the Nazis. Laudable in itself, the gesture was marred by the fact that it was donated by oil magnate Joseph H. Hazen, suggesting

II. The Symposium

that it was as much an expression of the "brain-drain" that has robbed the rest of the world of so much of its its scientific talent, as it was a tribute to the IAS, created by Abraham Flexner in the 30s for refugee scientists from Europe. (This may be considered almost quibbling, and we are indeed grateful that the IAS provided a home for Einstein, Gödel, Weyl and others that the world could ill afford to abandon.)

(2) A "gift" of $700,000 from the West German government to the IAS to support one senior professorship and two fellowships for a period of 5 years. In context this merely reinforced the sorry fact that the delegations to the ECS did not include a single person living in the Third World. When is the last time that West Germany or the US put aside $700,000 to promote the development of physics in India or Somalia?

(3) The presence on the stage, with a single exception, of bankers (Joseph Hazen, J. Richardson Dilworth, Howard Petersen) and highly placed establishment bureaucrats (Harry Woolf, Phillip Handler of the National Academy of Sciences and Jürgen Schmüde, Federal Minister of Education and Science for West Germany).

The lone exception was the Italian mathematician and IAS fellow Tullio Regge. He was waiting to receive the 1979 Einstein medal and remained silent all through the proceedings, not even giving an acceptance speech at the award ceremony itself.

The omission was the systematic exclusion throughout the Einstein Centennial Symposium of any representative from Switzerland, Einstein's adopted homeland. Einstein became a Swiss citizen in 1901. He was still one at the time of his death. All of his higher education was at Swiss universities. His revolutionary papers in Thermodynamics and Relativity were written in Zurich. During a break between lectures on Tuesday afternoon I got into a conversation with a Swiss government official in charge of scientific research, Roland Hartmann. He was present at the ECS "without portfolio." He let me know that he was "profoundly shocked" at this obviously politically motivated snub against his country. Einstein was both a Swiss citizen and a naturalized American, one of the very few persons in the world to have a true double nationality.

Herr Hartmann was therefore in a much better position than I to reach essentially the same conclusion: that the organizers of the event had considered the overt politics between the scientific establishments of West Germany and the United States as more important than any sincere homage to the memory of Albert Einstein. Yet judging from the tumultuous applause of the audience seated in the large auditorium of the Henry Chauncey Conference Center there must have been more than one individual in it who imagined that they were participating in a sincere commemorative to Einstein!

Some Strangeness in the Proportion

All this seemed incomprehensible to me; one might say that there was some strangeness in the proportion. A mere 40-year old at the time, I was still young and innocent. The impression I took away with me from the Opening Ceremonies was that of a captive witness at a reunion of capitalist old boy's clubs who, under the cloak of grand and fatuous gestures, were busily buying and selling the grey matter of the Trilateral Commission.

II. The Symposium

Now and then one of these aged plutocrats would deflate his rhetoric to toss a bone to the gaping throngs, such as a "reparations check" designed to redeem the unmentionable horrors of the past. Some soothing verbiage about the "nascent maturity" of American science was provided by Harry Woolf as a way of glossing over the unmentionable horrors of the present.

Oh well! There is a quality of obscenity to all staged historic events. My embarrassment was my own business. That I turned my chair at right angles to the stage and looked out the window through the whole of Jürgen Schmüde's presentation speech concerned no-one but myself. I am very much in the camp of those who believe we need to make every effort to absolve the Germany of the present from the Germany of the past. Our own recent history from Vietnam to Central America supplies the invidious comparisons.

What turned me into a one-man wet blanket at a time when it was expected that everyone would be having fun, was the mindless accord from all the "big minds" to corny political gestures that could serve no other purpose than that of lending an intellectual veneer to the merchandising between scientific, political, economic and military elites, temporarily united through the accidents of history against the Communist and the Third World.

A survey of the delegations to the Symposium lends further support to the views expressed above concerning its overt politics, while also saying something about its covert politics. The number of invited delegates was 120. Twenty-five, or about 22%, were currently resident at the IAS. Princeton University faculty contributed another 12%. Thus the "provincial sector" amounted to 1/3rd of the total. Many of the IAS fellows were foreign scholars; still, the vaunted "internationalism" of the ECS was a convenient label only. The number of delegates then living in the US was 100, or 90% of the total. To be fair, among these were famous refugees who'd been driven out of Europe by WWII and the Soviet takeover of Eastern Europe: Bergmann, Bargmann, Pais, Wigner....

Picking up the statistics from the other end, the number of invited women scientists was about 3%, a mere handful, less than even tokenism would dictate. Furthermore, most of these (there were after all, only 4!) did not bother to attend: most of 4 means 3, which leaves few indeed, that is to say the prominent physicist Madame Wen Chung-Wu (listed in neither the catalog of contributors nor the table of contents of *Some Strangeness...*). Notably absent also was Bruria Kauffman, whom I would have liked to have met again after so many years.

The only other women officially in attendance were nonscientists closely associated with Einstein in his lifetime: Helen Dukas, Einstein's secretary for 50 years, Mrs. Elizabeth M. Einstein, his daughter-in-law, and Ilse Rosenthal-Schneider, an old acquaintance from his Berlin days.

Following a familiar pattern, the two Russian delegates were unable to attend. It is customary to invite many more Russian scientists than one expects to show up. The Soviet Union has nothing comparable to the large conferences in the West which, despite their failings, are of great importance in the development of modern science. Russian scientists line up years in advance to attend conferences outside the country. Permission to attend, when granted, is more often than not withdrawn at the last minute. As a replacement Russia sent us a reporter from Izvestia. He spent much of the time drinking, and was not notably better prepared for the job than the majority of his American counterparts.

The tiny number of delegates from the Third World were all resident in the United States or had obtained

II. The Symposium

American citizenship: Claudio Teitelboim (Chile) Chandrasekhar (India), S.S. Chern (China). Here is the distribution by country of residence:

United States	100
England	5
Canada	1
West Germany	3
Austria	1
Switzerland	1
Israel	1
Italy	1
France	1
China	2
Japan	1
Russia (absent)	2

What is displayed here is a gazetteer of the affluent world as it might be compiled by some acutely myopic banker! (Some of the national contingents were larger than indicated because some of their members were working at jobs in the U.S. Thus Sciama, Dirac and Penrose, listed as being in the US, should actually be considered part of the English block at the Symposium.)

The number of delegates not incarcerated in a university was less than 4%, no more than 5. The academic overkill of the conference was predictable, yet depressing when it is remembered than the man it was designed to honor did his ground-breaking initial research as a Swiss patent clerk, 3rd Class, for 7 years. (Recently, when I pointed this out to an editor from Physics Reviews he said: "Well, things are different these days." Are they?)

The more one ponders these statistics, the further away one gets from the sense that the ECS had very much to do with respect for Albert Einstein. Einstein was a good friend of Marie Curie, yet today's Marie Curies were not on the guest list. Einstein worked for almost a decade at the lowest rung of the scientific establishment, an undistinguished, poorly paid civil servant, yet only men of the highest academic distinction were among the invitees. Einstein was a friend of Arnold Schönberg in Berlin and later in Princeton (there is a famous picture of them together with Leopold Godowsky), another bold rebel in his own field. Yet only the overly performed (though gorgeous) quartets of Mozart and Beethoven were permitted to disturb the somnabulators present at the two scheduled concerts by the Juilliard Quartet. (I'm very curious to know what Albert Einstein thought of Schönberg's music. There's no way of second-guessing a mind at that level.)

Einstein, Where Were You?

Finally: although the general public knows about Einstein as the man whose discoveries made the A-bomb possible, the whole subject of the nuclear age was categorically excluded from the podium. All of the talks with titles such as "Einstein, the Man," "Einstein in History," "Reminiscences of Einstein," and the like, were all too obviously distractions designed to side-step the real issues. I attended these of course, as I attended everything. The speakers at these talks warbled their ariosos about Einstein's humanitarianism, his social concerns, his simple, unassuming nature. One could not have divined from such tame eulogies that Albert Einstein was a principled and dedicated left-wing radical all of his life who was closely watched by the police in Kaiser Wilhelm's Germany in WWI, and almost deported by McCarthy in the 50s! A much better picture of the political man appears in "The Einstein File" and in "Einstein in Love."

II. The Symposium

Einstein—where were you? Was it to witness this that I had wandered the roads to Princeton, a penniless pilgrim? Was this the appropriate event to honor the life that made you the living metaphor of your millenially homeless people, of the intellectual torn from his roots, of the sage indifferent to titles and honors? Was praise of humanism without the substance of humanity, the masking of political horrors by political gestures, the kindling of academic rivalries, the crude evidence of sexist, economic and nationalist prejudices a fitting memorial to your unique life?

Yet after all is said and done Einstein was there. Of course; I had not erred in coming. He was present in many ways, though not always as he would have wished. He was present in the invigorating discussion of Relativity, Quantum Theory and Cosmology. He emerged from time to time in some of the unguarded anecdotes told about him by persons who'd worked with him. Certainly he was there in the magnificent Jacques Lipschitz sculpture "Arrival" that was given a prominent place in the cafeteria on the lower level, its powerful message indifferent to the uses of men and institutions. And in the works of Mozart and Beethoven as well. Most scientists cannot find the patience to listen to a contemporary music that so accurately reflects the culture science has done so much to create; but this does not depreciate the awe-inspiring intellectual and spiritual vision of these 18th century paradigms, that will outlast all changes of fad and fashion.

Einstein's habits, his spiritual struggles, his beliefs and the actions he took on their behalf, all these things were present in varying degrees. Yet in the elaborate rituals performed in his honor the man somehow fell through the cracks. The audiences listened to natural philosophers with no philosophy of nature; to humanists without humanity; to the outpourings of large imaginations shackled to academic institutions; to credentials riding hard over merit;

to politics prevailing over sincerity—everything that had revolted Einstein in his lifetime.

Yet despite these bureaucratic conspiracies the spirit of Einstein could not be stifled. A living light emerged from these 5 days of the Einstein Centennial Symposium, the searing flame of scientific inquiry, the offering of the latest fruits, of seeds sown by a life dedicated to justice and truth. That which had been denied admission to the Centennial Symposium came in through the back door, like the beggar who, though he be the guest of honor, is turned away from the feast yet comes in through the kitchen.

There was much that was positive at the Symposium, and many things to be learned: modern science will ever be proof against the scientists. Let us keep alive the fervent hope that, seen from the vantage of future generations, the historical illiteracy of today will have little effect, and all that will be retained is in the realms of speculative light.

Blueprint for a Symposium

Apart from the inevitable window-dressing, the Einstein Centennial Symposium consisted of 28 sessions spread over a period of 4 days. These ran consecutively in a single auditorium, the Henry Chauncey Conference Center, and not, as with many large conferences, concurrently in 3 or 4 auditoriums. Given that this auditorium is adjacent to the dining rooms, one could have, if so desired, stayed in the same building from 9 AM until midnight; a good approximation to my actual schedule.

The format governing all of the sessions was identical: 2 talks, followed by a commentary, after which the floor was opened for general discussion. The so-called "commentaries" were in fact independent papers, relating (if at all)

II. The Symposium

to the talks on which they were presumed to comment by a vague homology of subject matter.

This may have irked some persons in the audience, yet as someone who'd been away from the field for so long, I was gratified to be able to take in 24 original papers a day rather than 16. I don't imagine that the speakers minded the fact that their paper was serving as the *basso ostinato* for an original Passacaglia rather than as an aria to be ornamented.

The level of interest was always high even as the subject matter was bewildering in its variety. A loose overall structure grouped the lectures around general themes and related fields. Monday's sessions were for the most part devoted to the history of the development of Relativity and Quantum Theory in the early decades of the 20th century. A few talks about applications of relativity theory to engineering somehow wound up being included in the schedule.

Tuesday's fare was extremely diverse, as Relativity, Special and General, were examined from many points of view: theoretical, experimental, historical, mathematical.

The most exciting part of the Symposium was concentrated on Wednesday: 8 survey papers on recent developments in Cosmology by the leading figures in that subject. The self-conscious organizers of the ECS had ingenuously labeled Wednesday morning's seminars *The Universe*; those in the afternoon were called *The Universe, Continued*! Did anyone other than myself suspect that we might all be characters in a bimodal Borges fable?

Thursday was the last day of the ECS, and was divided into three parts.

(1) The talks in the morning dealt with current research on Unified Field Theories, Quantum Gravity and Super Gravity. To us mathematicians they seemed like

a dream come true: The universe really is an object of higher mathematics! To your John Q. Public it would have sounded like the kind of arcane mystification conjured up by the mere mention of the word "Relativity." It should be noted that the papers delivered at the Symposium were not technically demanding: a BSc in math or physics would have sufficed to understand most of them. The exceptions to this rule were the papers on Quantum Gravity and Super Gravity, and the paper of Tullio Regge using the language of differential forms.

(2) Thursday afternoon was given over to personal tributes to Einstein by the tribal elders. This aspect of the Symposium properly belongs to Comparative Religion, and will be discussed as such.

(3) Finally that evening there were the Closing Ceremonies. As mentioned before I and most of the others skipped these.

Rather than following a strict chronological order, my report has been arranged in terms of subject matter into 4 categories which are not mutually exclusive. Anecdotes and human interest stories tied to specific days have been interspersed to suggest the chronology. The 4 categories are:

(1) The Einstein Age, 1900–1926

(2) Experimental Relativity

(3) Cosmology

(4) Quantum Gravity; Super Gravity.

II. The Symposium

The Einstein Age: 1900–1926

The opening lecture of historian Gerald Holton on Monday morning established an image of Einstein as a young man that would echo in our minds throughout the Symposium. It is more than a truism to recall that Einstein's roots lay in the 19th century. His methodology, his way of thinking about scientific matters, his ambitions were recognizably those of the previous century. This was a individualist in the post-romantic tradition, working, if not totally in isolation, yet alone, applying techniques and concepts invented by himself. How different from the world of physics today, with its large research crews, expensive equipment and stifling bureaucracy! Holton's Einstein was untrammeled, poetic, imaginative rather than painstaking, a creative genius whose findings belied the narrow positivism of his most frequently acknowledged mentor, Ernst Mach.

Yet: although theory might well take precedence over the mere accumulation of data, it was all the same rigorous theory, its principles properly arranged, axioms above postulates, postulates above data, data above predictions. When no principles were to be found, he invented them (*Principle of Relativity*, *Light Principle*, *Principle of Equivalence*, *Mach's Principle*). A notion of seemingly little importance might suddenly find itself raised up to a law of nature (*the identity of gravitational and inertial mass*). Then again a rotted lichen, bloated with cancer, would be sheared off the tree of knowledge (*The ether concept*). Sometimes a basic axiom without which physics as a science is inconceivable would be boldly rescued from its detractors (*conservation of energy*); while other ideas fundamental to our way of thinking were conveniently dropped (*simultaneity*). Intuitions about the "rightness" of things (*"The good Lord would never allow...*") could be deemed more important than the supposed "hardness" of facts. Indeed, a finely tuned judgment was often needed to decide which facts were "harder" than others!

The biography of Einstein's early years reads more like those of the writers and composers of the 19th century rather than of the scientists of either the 19th or the 20th: alienation from schools (starting with the Gymnasium); years of unrecognized labor; perhaps a few legends of precocious genius thrown it from hindsight; little indebtedness to teachers or colleagues; a penchant for solitary meditation concealing a volcanic ferment.

All of the historians who spoke about Einstein's early papers (Gerald Holton, Martin Klein, John Stachel, Abraham Pais) were unanimous in their opinion that their many concerns could be traced to one crucial philosophical uneasiness: *thermodynamics, the theory of heat, was statistical in theory but not in the natural world.*

The Maxwell-Boltzmann probability distribution, designed to give a mathematical description of the long-range tendency for all hot things to cool, was not supported by any experimental evidence for the existence of random variations or "fluctuations." Einstein's preference for strict causation (which did not prevent him from employing probability and statistics like a master), required that the experimental evidence for a non-statistical phenomenon be placed on a non-statistical causal basis. From the onset of his career Albert Einstein was as much philosopher as physicist.

This personal preoccupation (for it appears that few other physicists were troubled by such matters) led Einstein, via a trajectory of 9 ground-breaking papers on thermodynamics, to that interface where fluctuations were most likely to occur, the interchanges between electromagnetic and mechanical energy. Together with Max Planck,

II. The Symposium

Einstein would become intrigued by that interchange of radiant energy with molecular vibration which goes by the name of *black-body radiation*, the phenomenon by which heated objects glow with different colors at different temperatures. This was the root anomaly that led Planck directly to the Quantum Theory.

In the course of writing these papers Einstein's horizons widened. Yet his methods throughout his life remained much the same. We find him responding with an almost visceral uneasiness when dealing with defective theories, not because there was anything wrong with their underlying concepts, but because he sensed they'd been shuffled in the wrong order. He distrusted *ad hoc* notions, arbitrary entities invented to explain the contradictions between theory and experiment: like the universal all-permeating ether, invisible yet infinitely rigid, required by the Huyghens-Maxwell wave theory of light. Such notions always introduce new problems, leading to more contradictions which require yet another *ad hoc* entity, in an unending process.

In 1895 for example, H.A. Lorentz introduced length contractions into electrodynamics in a last-ditch stand to explain why the ether was undetectable. It was being suggested by some that one ought to separate cause and effect in space and time. Even the most fundamental principle of them all, the conservation of energy, was in danger of being discarded. Accurate formulae were being derived by faulty reasoning from incorrect assumptions. (Mathematicians who habitually deride physics claim that the whole subject is like that.) The epic tale of the chaos into which theoretical physics had descended at the turn of the century has been retold many times. To quote the refrain physicists like to chant, there was "no firm foundation."

Max Planck Defenestrated; Wigner to the Rescue

Each of the guest lecturers in turn spoke of the lucidity, power and sure intuition of Einstein's intellectual vision in the confusion of the times. Thomas Kuhn, the MIT philosopher of science known for his neo-Hegelian musings on the dialectical process governing the life and death of paradigms, had compiled a thick scrapbook of historical citations to convince the Symposium that Einstein probably deserved more credit for Planck's quantum hypothesis than did Planck himself. It wasn't a pet theory; both Martin Klein and Abraham Pais gave it their full support.

In the discussions that followed, the aged Eugene Wigner arose somewhat ponderously from his seat in the front row to put Max Planck back onto the pedestal from which he was about to be rudely toppled. One cannot, he said, understand the thinking of the major figures in the field without being an eyewitness to the confusion that reigned in physics at the turn of the century. Equations that worked, proposed in defiance of all theory, sometimes on the basis of sheer nonsense, were being cobbled together from empirical data by researchers at every level. Max Planck's somewhat confused rationale for combining the incompatible laws used to describe black-body radiation, (Wien's Law and the Rayleigh-Jeans Law), constituted as much of an advance over their work, as Einstein's subsequent clarification of the quantum hypothesis did over Planck's.

II. The Symposium

Albert Einstein Defenestrated; Dirac to the Rescue

Over the course of the first day of the Symposium, Albert Einstein himself came in for his share of the debunking. Arthur I. Miller flashed slides of photographs of a recently uncovered cache of letters between H.A. Lorentz and the famous French mathematician Henri Poincaré. These demonstrate that Poincaré had created all of the mathematics and much of the vocabulary of Special Relativity (including the word "relativity"!) before Einstein had written any of his papers on the subject.

And indeed there is a real question as to whether Poincaré or Einstein ought to be credited as the father of Relativity. When I studied mathematical physics for a term at the *Institut Poincaré* in Paris in 1968, my instructors taught that Einstein should not be given credit for Special Relativity. Einstein himself is quoted as saying that he believed that Paul Langevin would have developed Special Relativity in the long run, but no-one but himself could have invented General Relativity.

Arthur Miller's historic reconstruction was skillfully deployed. He brought to life many influential figures of that period now almost unknown to the scientific world: Max Abraham, Paul Langevin, Walter Kaufmann, Michele Angelo Besso. By the end of his lecture we were doubting that Einstein deserved any credit at all for the theory now inextricably associated with his name. It remained for the chairman of the historical sessions, one of the world's greatest mathematical physicists, to give the final word: Paul Dirac.

The sensation produced that Monday morning by this eminent elder statesman of science rising in the defense of his great predecessor, can only be compared to the imagined presence of Aristotle at a conference in honor of Socrates being held at the Platonic Academy. White-haired and frail, still driven by enormous energy, his extraordinary intelligence stamped on his features like a medallion that has broken the mold, he bore himself with a greater degree of personal independence than one would normally expect from a pre-canonized scientific immortal. With a small number of apt comments, Dirac summed up half a dozen talks and several hours of dialogue. As is the way with great scientists, his observations went to the heart of the matter unencumbered with detail.

Einstein, he reminded us, had taken the Lorentzian model, which was restricted to electrons alone, and extended it to all physical phenomena. By doing so he was able to predict an *immense range of effects that had never been seen and that no-one would have ever thought to look for.* Such a feat of intellectual audacity was almost unique in history.

Through the postulation of a new universal law, the constancy of the speed of light in all reference frames, he kicked away the last vestiges of the ether hypothesis. Dirac then went on to say that a measure of the sureness of Einstein's intuition can be found in recent findings that indicate that he was partly wrong. The *microwave background radiation*—thought to be the dying echo of the Big Bang 14 billion years ago, a radio static arriving from all parts of the sky, with a very low frequency, in terms of its equivalence in heat, of about 3° above Absolute Zero—functions as a kind of fixed reference frame, or ether. This resurrected form of the ether hypothesis is so totally unlike the naïve conceptions of the 19th century, that it must be treated as a distinct notion. It could never have been discovered had not its earlier incarnation been abandoned.

To my mind, Professor Dirac's paraphrase summed up all the historical debates delivered that morning. The historical studies that followed had little to say about Einstein's relationship to his times, or even to the world of science. As I was to discover, an unreasonable amount of Symposium's time would be given over to dusting off the deathless null controversy of modern physics: *would General Relativity have been invented if Einstein hadn't done so?*

It was only after I began interpreting the Einstein Centennial Symposium as a quasi-religious gathering of the tribe around the camp-fires, that I was able to understand why so many precious hours were being wasted on this meaningless scholastic debate. This was the moment for the shamans to come together and recite the magical conundrums dispensed to Mankind by the Gods at the beginning of the world.

Einstein the myth, not the man, demanded its own performance of rites and rituals, one of the oldest of which is the re-enactment of the miraculous deeds of prowess of the incarnated epic hero. Even as Gilgamesh went into the Land of the Living to set up the names of the Gods, so did the Mythic Architect build his temple of General Relativity in the Citadel of Knowledge.

Dinner with Subramanyan Chandrasekhar

Monday Evening. Dinner: I am seated at a table in the IAS cafeteria. To my left sits Dr. Isaacson of the National Science Foundation. Opposite me are Subramanyan Chandrasekhar and his wife, Lalitha. Her features are twisted with the customary boredom of academic spouses at scientific conferences on subjects about which they know nothing and care little. I am calling her a "spouse" because the very next day Helen Dukas explained to me that the technical term for a non-scientist mate has become "spouse." Can it be that feminism is beginning to crash the gates of this most indomitable of masculine domains, the physics community?

As I sit there, silenced and ignored, I listen mouth agape as Isaacson and Chandrasekhar cynically trade recipes for getting rid of candidates for funding whose projects have been rejected.

Lunch With the Bureaucrats

Tuesday Afternoon: Lunch with William Dillon of the Smithsonian Museum, Bill Wing of the IAS, and reporters from United Press International and the Washington Post. Dirac sits at the far end of the same table, surrounded by other scientists and deep in private conversation.

William Dillon wants us to help him with a problem he's been trying to solve: he is in charge of putting together the next Smithsonian exhibition: a panorama of Life, from the bogs of pre-Cambria to the present day. "There are no new ideas," he sighs, "It's been done so often before."

One of the British newsmen and myself begin suggesting ways of introducing themes from modern genetics and biochemistry: "Don't forget DNA," the reporter says, "There was a show about Crick and Watson in London last year."

We derive a small, pardonable satisfaction in confusing Dillon even more than before. The conversation then drifts into the usual generalizations about the inability of scientists to relate their ideas to the public at large. Somebody mentioned Michael Rockefeller's film about New Guinea: here was a scientist who'd gotten so close to ordinary people that they allowed him to film the most intimate details of their daily round. (As we know, it cost him his life.)

Dillon's comment: "He wasn't a real Rockefeller. He was probably the only non-WASP among them."

This prompted Bill Wing to ask: "How would you characterize Nelson Rockefeller?" It is Mitchell of the Washington Post who has the last word: "A hornet."

Experimental Relativity: Irwin Shapiro

It was from the sessions on experimental relativity that I learned (to my complete satisfaction) the answer to the question I'd posed to Harry Woolf. The theory of relativity has entered the Space Age. It is most appropriate to the Einstein Year of 1979 that the technology required for detecting the exceedingly feeble 1st and 2nd order effects of General Relativity has finally caught up with the predictions of theory.

The instruments of the famous Eddington eclipse expedition of 1919 appear by comparison to have been held together with Scotch tape and paper clips. On that occasion teams of poorly funded astronomers, toting little more than what we today might call box cameras, traveled to Brazil and West Africa to find evidence for the bending of light in a gravitational field. There amounted to two serviceable images, only one of which could be successfully exploited by Arthur Eddington (through a laudable albeit notorious violation of strict scientific method) to announce the triumph of General Relativity.

Stepping out on the brink of the 80s, we were given previews of magnificent space operas involving at least 24 space craft, two of them circling, and two more stationed on, Mars. Radio telescopes around the world were already poised to monitor the cross-hatched radio signals they were expected to beam back to astronomers on earth.

We were shown previews of the Jupiter and Venus probes, already launched and slated to occult later this year; this is a new method for measuring the deflection of light by gravitational fields to a very high degree of precision. Methods employing occultation in space eliminate atmospheric interference in signal transmission. Many new directions are opening up for the detection of gravity waves. They are more than just a prediction of General Relativity, they are intrinsic to the shape of the theory: the tensor equations that equate gravitation with the bending of space-time are hydrodynamic in character. Space-time actually behaves like a liquid in which one expects to find both waves and turbulence.

Yet to date gravity wave-fronts haven't been detected. However Irwin Shapiro displayed some very interesting calculations by John Taylor which indicate that this situation may be about to change. Peaks of radio emissions from pulsars circling each other in binary pairs have been statistically analyzed. The graph of a faint parabolic decay in these emissions is show in the paper of Taylor, Fowler and McCulloch (*Nature*, 277, 1979). Shapiro interpreted this as evidence for gravity waves.

Taylor's graph grows by about 2 points a year. It seems a bit far-fetched to deduce the presence of a phenomenon as universal as gravitational waves, on the basis of 20 points in 12 years. However, when I visited with Phillip Morrison at MIT in 1993, he said that Taylor's parabola continues to follow its predicted trajectory year after year. He was convinced by the evidence. I wouldn't dream of contesting the opinions of Philip Morrison in his own field. As for Black Holes, he was still stating at that late date, "I'll be

II. The Symposium

convinced when I see one." (A joke. Regrettably, Dr. Morrison died a few weeks before this revision of the article was initiated in April, 2005.)

This report presents only a small selection of the advances in experimental relativity discussed at the Symposium. The "symphonic scores" of these experiments are as elaborate as any contemporary piece of Boulez or Stockhausen. Most of the GR experiments are "piggy-backed" onto the large projects of exploration of the Space Program, and involve little extra technical work from the ground crews. Therefore their cost is minute in comparison to the Himalayan fiscal landscape of NASA.

The opportunity to be able to employ the sophisticated technology of orbiting space laboratories is balanced by a number of serious disadvantages. Owing to the expenses and security risks involved in routine NASA launches, the scientists who design the experimental GR packages are not allowed to be anywhere near their own equipment before it goes into launch. This means that it is impossible to eliminate or even reduce systemic errors.

It was therefore all the more gratifying to learn that all of the GR experiments performed in outer space from 1974 to 1979 have shown "a remarkable verification of the predictions of the theory." One can think of no more fitting birthday present to late Albert Einstein, than this simple comment of Irwin Shapiro's at the end of his talk.

The Cocktail Party

Wednesday Evening: The Henry Chauncey Conference Center has been transformed into a cocktail lounge. Long tables covered with blue tablecloths and stacked with

The Cocktail Party

glasses and bottles line the gangways leading from the auditorium to the lobby of the Annex.

I was early coming into the Annex. At most a dozen persons were standing about cradling pre-dinner *apéritifs*. Twenty minutes later the auditorium is packed solid. A rapid stream of noisy clicks can be heard coming from the "fame-density" meters!

I order a Scotch on the rocks from a young bartender. He asks me to explain to him the exact meaning of the terms "lightyears" and "parsecs." He's never "shown much aptitude for math," he explained, and dropped out of school to go into the army. I discover that my ability to explain such subject matter has improved through attendance at the conference. He returns to his work, I float back into the auditorium.

A young man, down-at-the-heels and evidently out of place, holds a drink and is leaning against a bare cinderblock pillar. Perhaps it is because I, too, am down-at-the-heels that I strike up a conversation with him. With nervousness masked by bravado he announces that he is deliberately crashing the cocktail party and hopes to get a free meal out of it as well:

"These people claim to be civilized," he sneers. "If they were really as enlightened as they claim to be, they wouldn't hesitate to open their dining-rooms to hungry scholars." I offer that he ought be making a distinction between the administrators of the Institute, its directors and trustees, and the scientific delegations, a distinction that I myself have trouble believing in. My deeper feeling is that the scientists are probably turned off by him, not because they want to deprive him of a meal, but because of his simple-minded, let us not say stupid, interpretation of the situation. I wish him well and move on.

Sitting down for dinner, I find myself in the company of the members of film crew. On the spot they give me a

II. The Symposium

job: to point out the real celebrities in this motley crowd. (Note: it has been claimed that T.S. Eliot's play "The Cocktail Party" is based on his stay at the IAS.)

Social Responsibility at the IAS

Wednesday Afternoon: I'm sitting on a couch, alone in the large reading room of Fuld Hall, the principal building of the Institute for Advanced Study. Dominic, a security guard, comes over to share a few words with me:

"Take a look at those magazines, will you?" I stand up and walk over to the magazine rack: *Time, Newsweek, US News and World Report, Punch.* "Someone asked the library to order this one. It started coming a few weeks ago." Dominic pulls out an issue of *The National Review* and passes it over to me.

"It's more than I can understand," he continues, "When I took this job I thought the place was filled with political radicals. Apparently having brains doesn't mean you know anything!"

Indeed, the radical fringe at the Institute cannot be very large. (Einstein would have belonged to it.) No issues of *Mother Jones; Seven Days; The Nation; Le Nouvel Observateur; Akwesasne Notes.* Once again, Dominic asks me to help him understand some perplexing idea of modern physics. I launch into a discussion but am unable to continue: his supervisor has suddenly appeared in the doorway to the lounge and regards him with a certain amount of ill-humor. Dominic excuses himself and hurries back to his work.

Cosmology

Sciama's Visit to Cygnus X-1

Irwin Shapiro survey of experimental relativity was finished. The moment the discussion period opened the stocky English cosmologist Dennis Sciama rose up out of a chair near the front of the auditorium. Waving his arms he uttered the "good news" in the apocalyptic tones befitting his subject. The affinities of cosmology to grand opera derive from their common goals:

(1) To reveal, beneath the raw data of experience, the eternal invariants that govern the cosmos, or the eternal truths of love and fate.

(2) To provide simple, gratifying answers to deep questions about our place in the universe and the direction in which we are headed.

Providentially for some, cosmology is a subject in which almost all relevant information is either inaccessible or unknowable. Given this state of affairs (similar to what one finds in hominid paleontology) its practitioners tend to make sweeping generalizations on the basis of a few scattered observations, with only a minor concern for confirmation through prediction. It's impossible to perform

III. Cosmology

cosmological experiments since there's not much that is predictable. (The Microwave Background Radiation is a notable exception.)

Thus, although there are 8 commonly accepted solutions of the Einstein Field Equations, it is exceedingly difficult to imagine experiments that would decide between them. Cosmology is therefore divided between Observation and Theory, with little activity in the domain of Prediction. Karl Popper would probably have concluded that Cosmology is "insufficiently falsifiable" to qualify as a science.

These inherent limitations tend to put cosmologists on the defensive. Since they lack the data needed to back up their hypotheses they are known to take cover behind the argument that the data is also insufficient to prove them wrong. Add to this the indisputable fact that their more earthbound (when not hidebound) colleagues, the physicists, have a tendency to cavort about with superior notions of the "purity" of their science, and the cosmologists run the risk of falling foul of the perennial witch-hunts mounted against the so-called "pseudo-sciences" (normally the social sciences, but sometimes hapless "reconstructive sciences" like geology and cosmology). Let a cosmologist make a slip of the tongue and people will start calling him an astrologer behind his back—worse than a truck driver! This snobbery merely increases the cosmologist's tendency to take refuge in dogmatism.

I hope that my thumbnail sketch of the science has put us in a better position to understand why Dr. Sciama accompanied his pronouncements with so many rhetorical flourishes. It also explains the mixture of weary skepticism and cautious interest they aroused.

Sciama told us that the hitherto unclassifiable object at the center of the Galaxy Cygnus X-1, a pathological radio wave and X-ray source, is indeed a Black Hole. He presented no new findings to back up this assertion. In my conversations with several delegates afterwards, no-one would commit himself beyond stating that the evidence for radiation coming from the galaxy is consistent with the presence of a Black Hole. (Since then, many candidates for Black Holes have been found, including a few at the center of our own galaxy. The evidence is strong though not conclusive: there is still something hypothetical in the very existence of Black Holes.)

Sciama then went on to say that because the object at the center of Cygnus X-1 is a Black Hole, the first- and second-order effects of General Relativity (normally so faint that they are barely detectable even with the most advanced technology) will be biblical in grandeur. Orbital shifts, like the shift in the perihelion of Mercury, the 43" of arc that gave GR its initial boost, should be of the order of 100 degrees or more, almost one-third of a complete rotation. Gravity waves, to date undetected, will be visible in gargantuan vortices and turbulence.

Sciama's final announcement was that investigations were underway for the evidence of a "dual world" inside Cygnus X-1, in which the properties of space and time would be interchanged!

All of these revelations were very exciting, and if they hadn't been presented with such chiliastic eloquence they might have been even more so. One way or another, Sciama's thunderous Toccata and Fugue launched the scientifically technical part of the ECS.

The Battle of Princeton

Wednesday, March 17: The history books will remember this date as the day on which the British returned to replay

III. Cosmology

the Battle of Princeton. One might describe it as a kind of Norman Conquest in reverse, with British cosmologists combining forces from strategic locations in England and North America to wrest the trophies they'd already won.

In force were, among others, Hawking, Rees, Sciama, Penrose, Dirac, emerging from the residual Background Radiation of Bondi, Gold, Hoyle, Lovell, Whitrow, Eddington, Jeans... going back to the Herschels, even to Newton himself. The mere fact of their domination of the Einstein Centennial Symposium was deeply moving for anyone with some perspective on the history of science. It was the spectacle of the Newtonian heritage disbursing its cornucopia of fruits at the Einsteinian banquet.

This was the long-awaited vindication of the idiosyncratic, much ridiculed, characteristically stubborn tradition of British mathematics and astronomy, its obsession with geometry, it fondness for cumbersome notation, its inexhaustible fertility in theoretical physics (Newton, Maxwell, Rayleigh, Kelvin), its unapologetic Pythagoreanism (Eddington, Clifford, Dirac), its 300-year long commitment to the noblest of the "impractical" empirical sciences, astronomy. One had the sense that Newton's hands had stretched themselves across the centuries to join with our contemporaries, that one of the primary broken links in the European scientific tradition had finally been repaired.

The keystone to the British arch was Stephen Hawking, one of the towering minds in contemporary theoretical physics. In our science-drunk culture it is our tendency to regard towering minds in mathematics and physics as superior to towering minds in all other fields. It's possible that we might not think so highly of Einstein and Hawking if we believed that Oriental carpet weaving was the pinnacle of intellectual achievement. As for myself, I tend to share in the general consensus that exceptional ability in mathematics and physics does indicate the presence of a powerful mind. All the same I could not help but be amused overhearing a conversation between two of the photographers assigned to cover the event.

One said to the other: *"This place collects the greatest minds in the world!"* before crying out: *"Hey! Let's get going! That's one of the guys whose picture we've got to take!"* I was unable to distinguish the wizard to whom he was pointing from the crowd around him.

All of this is being said by way of preamble to the entrance of Stephen Hawking onto the stage of the ECS. In 1979 Hawking was still a young man in his 30s. Since 1965, as we know, he's been the victim of a progressively deteriorating form of multiple sclerosis. To the casual or ignorant gaze he would be considered a basket case. He sits in a wheelchair, unable to move, his head lolling to one side, constantly attended to by a team that rolls his chair, feeds him, and interprets the harsh rattles that gurgle through his throat in lieu of communication with the outside world. *(1979 was many years before the invention of the computer equipment that has since transformed his life.)* To the uninitiated his utterances are incomprehensible.

His interpreter for the Symposium was a Mr. Alan Lapedes. Lapedes is a somewhat forbidding individual, hardly what one might consider an extrovert. His traits must in fact have been of great value in keeping curiosity seekers at bay. My friends in Princeton told me that Lapedes, through a combination of personal affection and dedication to science, has sacrificed his own career to serve this stricken genius.

III. Cosmology

Upstairs, Downstairs at the IAS

Daily proximity with Hawking put me in mind of the compelling phrase of another sickness-tormented scientist, Blaise Pascal: *the greatness and misery of man.* This phrase maintained a constant presence in my meditations throughout the Symposium. As it turned out, a rather strange personal interaction took place that put it into a unique perspective:

During Tuesday's lunch-break I struck up a conversation with a pair of waitresses employed by the IAS. One of them was a college student working part-time. The other, a short woman in her middle 50s, had concluded, for some unfathomable reason, that I was more "approachable" than the other high dignitaries. To me she confessed to me that the "man in the wheelchair" had aroused her curiosity. She knew that he was a delegate to the Symposium because he wore a name tag. She couldn't imagine that a person in his condition could ever get to be a professor. When I explained to her that he was in fact one of the world's greatest astrophysicists, she was thoroughly dumbfounded.

She proceeded to explain her personal interest in the matter. She has a niece, now 18, who's been confined to nursing homes since the age of 5. Her illness resembles Hawking's. She was convinced that her niece had a high intelligence but didn't know the first thing about obtaining an education for her that would enable her to leave the asylum or, at least, live a constructive life within it.

I realized that she wanted me to arrange for her to speak to Hawking, so she could ask him for information about organizations that could provide the training and care that she felt her niece. She was much too inhibited to speak to so distinguished a person directly, and was asking me to serve as intermediary.

Inhibition, or let us rather say discretion, has never been my strong point, and I promised to look into the matter right away. Right after lunch I went into the Board Room, a private room for meetings and discussions in back of the dining-room, to look for Stephen Hawking. He was sitting at a long table at the back with Alan Lapedes, other citizens of the Cambridge clique and cameramen from the BBC. I'd concluded that the best tactic for approaching someone like Alan Lapedes was a pose of "strictly business." Lapedes did hear me out and conveyed my request. Hawking replied that he would be more than happy to meet with the waitress at any time that afternoon.

Returning to the kitchen I learned that she had left for the day. I therefore wrote her a note asking her to meet with me the next morning to set up a formal appointment. I then returned to the Board Room. Vaulting another Lepedes hurdle, I managed to obtain assurances that Hawking would arrange a time to see her the next day, or on any of the remaining days of the conference.

Around 10 the next morning, the elderly waitress was back at work setting out tables for the first coffee break. Stepping into the kitchen again to talk to her it became obvious that she'd begun to regret her rashness in asking me to set up the interview. There was a further cause for awkwardness in the impatient, somewhat unfriendly manner of a kitchen overseer who seemed to think that the delegates shouldn't be associating with the help. It appeared that the waitress felt very comfortable with this worldview. She couldn't meet with Hawking that day, she said, because of her work schedule. Before walking away, she conceded that she might be able to spare a few minutes during her lunch break.

III. Cosmology

A little before noon I once again entered the cafeteria and caught up with her sitting at a table surrounded by other waitresses. This time I faced an undeniable wall of hostility. She indicated in various ways that people like myself ought to be minding their own business. I did however manage to transmit the information that Hawking and Lapedes had indicated a willingness to talk with her at some time which was convenient for her, say after work or early in the morning. She ignored me, not even bothering to nod. I returned to the afternoon sessions of the Symposium.

Whenever I ran into her after that she went out of her way to avoid me. Social barriers exist everywhere of course, but the last time I'd seen them so rigidly enforced was in the job I'd taken on back in 1957 as busboy in a resort for the super-rich, the Otesaga Hotel in Cooperstown, NY. Needless to say, she never did follow through on her desire to meet with Hawking.

From the Cosmological Order to the Princeton Caste System: the greatness and misery of Man. The "cosmology" of human motivation far belittles the relatively innocuous cosmology of the heavens. Writing during another great age of science, Sophocles states, in the *Antigone*: "All things are strange, yet nothing stranger than Man."

Black Hole Phrenology

Many aspects of modern cosmology were touched on at the ECS. Here I relate only my impressions of talks given by Stephen Hawking and Martin J. Rees. Hawking's paper was read by Alan Lapedes standing beside Hawking's wheelchair on the stage. After summarizing recent developments in the field the talk transmitted, on a higher technical level, the substance of Hawking's article on the disintegration of Black Holes that appears in the *Scientific American* of February, 1977. The talk was organized around 3 topics:

1. The Cosmological Censorship Hypothesis;
2. The *No Hair* property;
3. The surface temperature of a Black Hole.

The Cosmological Censorship Hypothesis (CCH) is a classic "cover-up" with which the history of the sciences is replete: the ether, phlogiston, epicycles. These are used to shore up a theory in trouble by postulating entities which, by their very nature, cannot be observed. The CCH states that although a complete breakdown of causality occurs in the interior of a Black Hole, the universe outside the Hole remains causal, because this chaos (along with matter, strong and weak forces light and everything else) remains trapped within it under the force of gravity. The usefulness of this hypothesis lies in the fact that one can continue to use the equations of General Relativity to describe the "Large-Scale Structure of Space-Time" (the standard textbook on the subject by Hawking and Ellis)—*including the existence of Black Holes!*

The history of the sciences has shown that such *ad hoc* entities will eventually be discarded as unscientific, or, in the terminology of Karl Popper, as *metaphysical* rather than *physical*. As a holding action the "invisibility of chaos" may be useful, yet eventually it will be obliged to brave the light of day, if only to be given a shave by Occam. A Black Hole is said to "have no hair" because it is completely described by a finite number of parameters. Hawking cited 3: its mass, angular momentum and electric charge. However when the floor was open for discussion Claudio Teitelboim announced that he'd discovered a fourth parameter,

III. Cosmology

its internal spin. Strangely there is no mention of his announcement in *Some Strangeness in the Proportion*, where it does however appear indirectly in a statement by John Archibald Wheeler in his talk: *"there is another feature of the Black Hole, Claudio Teitelboim tells us, its spinor spin."*

The inclusion of internal spin does not rob the Black Hole of its meniscus of Absolute Perfection. Still I find it hard to believe that in this fatally flawed world there's anything that can get away with a paltry 4 parameters. A single atom requires over a hundred of quantum numbers to do it justice—and the list gets larger each year!

Objections were raised from the floor of the auditorium against the endowment of barren beauty to objects anyone has yet to see. According to the No Hair principle, the Black Hole ingests it surroundings whole, like a python devouring a pig. All of its complexity is absorbed and nothing returned: matter, antimatter, electrons, mesons, quanta, quarks, neutrinos…. Whatsoever be their strangeness in proportion previous to being ingested, they now serve only to fatten the 4 parameters.

Yet Hawking's own discovery, Hawking Radiation, coats the Black Hole with a tiny amount of hair! A sort of adolescent fuzz, nothing more. The Uncertainty Principle, the mainstay of Quantum Theory, sees to it that even the eternal glory of a Black Hole eventually undergoes disintegration.

Eternal Life and the Baryon Number

It was pointed out that this means that one abandon a fundamental principle of particle physics, the *Conservation of Baryon Number*: the quantitative difference between matter and anti-matter is an invariant in all interactions and processes of decay. Yuval Ne'eman asked if there might be some way of interpreting the equations to save baryon number conservation. Hawking's reply (through Lapedes):

> *"I find it interesting that people have such an emotional attachment to baryon conservation. This may be because most people do not believe in eternal life. They would like to hope that the particles which make up their bodies would live forever."*

The jest crystallizes the very style of contemporary physics. Scientists have been condemned throughout modern history for mocking the existence of a soul. Now it appears that they have as little use for matter!

The viability of the law of baryon conservation seems to depend upon whom one is talking with at a particular moment. The full quotation from Ne'eman reads:

> *"We owe a lot to baryon number. We owe our existence to the conservation of baryon number. Otherwise we would be floating in the universe as $E=mc^2$!"*

Fields and Particles

Today's physics community appears to be polarized into particle-ites and field-ites, with many shades of opinion falling between the extremes. The issues are quite difficult and involve as much philosophy as they do science. It's a pity that the physicists rarely bother to consult with the philosophers, though I do sympathize with their aversion to walking over to the ugliest building on campus, to climb 5 stories to a gloomy attic holding a few offices filled with

III. Cosmology

oversized furniture, unfriendly secretaries and virtually no social amenities, not even a lounge with free tea and coffee. The reason for this sad parsimony is simple: no-one's figured out a way to turn existentialism into a thermonuclear bomb.

As Abraham Pais pointed out in his engaging account of the birth of modern physics, there does not exist, even at this late date, any satisfactory definition, either ontological or epistemological, of the autonomous field in empty space.

If there were any place on earth to take a poll of the range of opinion in the field/particle debate, it would seem to be the ECS. At one extreme I found the physics professor from some SUNY campus who maintained that fields were nothing more than convenient devices for making calculations. At the other extreme stood Hawking himself, taking undisguised relish in tossing every principle of particle conservation onto the junk-heap of antiquated science and relying exclusively on entities implicit in the mathematics of field equations. One should also include the mathematicians at the Symposium, S.S. Chern and Tullio Regge, avid to replace everything physical, both fields and particles, by shopping lists of symmetry principles.

Quote from Tullio Regge:

> *"We must not forget, however, that physics, so to speak, is geometry plus an action principle."*

As it turned out, I socialized a bit with Tullio Regge under rather peculiar circumstances. Two newspaper photographers, knowing nothing about the event they'd been sent to cover, herded Regge and myself into the dining room on Thursday afternoon and posed us together as representative of the distinguished scientists at the Einstein Centennial Symposium!

Isaac Rabi Resuscitates the Media

The profundity, sophistication, and intellectual intensity of the field/particle debate (which shows no sign of cooling down after a century) should be contrasted with the manner in which it was viewed by the paparazzi responsible for covering it: on Thursday afternoon, March 18th, Isidor Isaac Rabi, Nobel prizewinner and crusty octogenarian, hoisted himself up painfully from his chair to tell the audience once more that only the things one sees in the laboratory are real. There was a certain amount of, well, I wouldn't call it demagoguery to his delivery, rather something in his tone of voice like that of a teacher scolding his recalcitrant students. To paraphrase his critique, the physics community must re-enter the laboratories to discover what one brings into it: the real reality that everyone really knows is really there. His is a perspective not too keen on distinguishing particles from fields.

One gets a glimpse into Rabi's world-view by the comment he once made to the effect that Oppenheimer's "failure" as a physicist was due to "overeducation" in the humanities:

> *"It seems to me that in some respects Oppenheimer was overeducated in those fields which lie outside the scientific tradition, such as his interest in religion, the Hindu religion in particular, which resulted in a feeling for the mystery of the universe which surrounded him almost like a fog...."*
>
> —*Who Got Einstein's Office?* pg. 147.

III. Cosmology

With the dying away of Rabi's final word, one saw a dozen journalists rise up out of their seats and dash like a coherent wave packet out of the auditorium to the Press Room. Curious myself to see what could have roused them after 5 days of collective indolence, I went back there myself shortly afterwards to see what all the fuss was about. I was astonished to find them all hard at work. One of them racked his brains to come up with a original headline of the genre: *Leading Scientist Cries Back To The Laboratories!* Another one was flipping the pages of the science encyclopedias to note down all of the distinctions and accolades bestowed upon the sage by a grateful humanity. A seasoned hack gave us the following advice: "Just write something like, *'He made advances in understanding the molecular structure of matter.'* You can always say that about a physicist."

From the entrails of a debate over the most challenging ontological dilemma of the 20th century, the "reality" of the mathematical constructs (fields, tensors, Schrödinger wave functions, space-time) which have undermined the "tangible" magnitudes (matter, time, space, momentum) that we are accustomed to encountering in daily life, the press corps had been roused from its slumbers once and only, when the perennial mugwump arose to pound his wooden ideas on the "real" floor on which he stood.

In the final analysis, the newsmen did not even take the time to appreciate what merit there was in Rabi's rebuke:

> "… I think Hilbert was once asked, in a certain mathematical colloquium, what he thought of a paper. And he said *'Kreide'* (chalk). And this is just a slight reminder that there is a real world."

The Cosmological Principle: Martin Rees is Miffed

With the delivery of the paper by Martin J. Rees, "The Size and Shape of the Universe," we move away from the Hawking Plenum of Gravitation to enter a more nebulous yet equally challenging Empyrean of Observational Cosmology.

For observational purposes the knowable universe is limited to the *photosphere*, the 14 billion or so light-years that a light signal travels from the instant of the Big Bang to earth-bound observers in the present. It is the cosmologist's version of Rabi's laboratory. Through a combination of the Postulate of Special Relativity which limits the transmission of signals to the speed of light, and the Hubble Law of the expansion of the galaxies, the photosphere falls short of the potential universe. (Note: Inflationary Theories and the hypothesis of dark energy have since put new wrinkles into this simplified model.)

The cosmologist enters this laboratory with a small number of ground rules which he considers essential to his science. Looking deep into the universe he is also looking back in time: light from a star one billion light years away from us takes one billion years to reach us. The astronomical scale is almost impossible to grasp in intuitive images: our Sun, dominating the daytime sky, is but a single star. Yet at a great distance a galaxy of a trillion stars is a speck of light invisible to the naked eye.

In order to organize the data coming in from a volume of such inconceivable magnitude, cosmology requires a postulate, suggested by the evidence, but which in fact is deemed necessary for the subject to qualify as a science:

III. Cosmology

The universe is homogeneous. In his follow-up commentary, Jim Peebles referred to it as the "Copernican argument":

> *"... that the view of the universe from most galaxies that seem to be equally good homes for observers would be quite different from our own galaxy ... seems unreasonable."*

One of the consequences of this assumption is that the universe that lies beyond the boundary of the photosphere, parts of which will become visible over billions of years, will have the same large-scale features as those that we can now see. And the same must have been true in the past, most of which is unknown.

The combination of the Homogeneity Principle with the Principle of Isotropy (that the universe looks the same in every direction) is commonly known as the Cosmological Principle. Quoting from Jim Peebles classic text *Principles of Physical Cosmology* (pg. 15):

> Milne's (1935) term, "Einstein's cosmological principle" is appropriate in the sense that the conditions of homogeneity and isotropy do greatly restrict the range of possible cosmologies, as Milne was among the first to appreciate.

and (pg. 16):

> Might the cosmological principle be elevated to a physical principle that has to be true? We should bear in mind that although some may be glad to accept the cosmological principle because it simplifies the mathematics, Einstein was motivated by something quite different:

The Cosmological Principle

the idea that a universe that is not homogeneous and isotropic in the large-scale average is absurd. Since the argument has proved successful, perhaps it is telling us something deep about the nature of the universe.

The Cosmological Principle fulfills the same function for Cosmology as the Uniformitarian Principle for geology, which allows one to reconstruct the past of the Earth from processes at work in the present. There isn't any way of "proving" such a principle, yet (following Lyell, the founder of the science) it is deemed necessary if geology is to stake its claim as a real science.

The problem is that there is a crucial difference between the metaphysics of geology and cosmology: one uses the Cosmological Principle to look backwards in time and uncovers the Big Bang! *Catastrophism*, the opposing view dear to organized religion that early geologists had to contend with, contends that the world we live in was formed by special moments of creation, such as the 7 days of Genesis and the Deluge.

Its correlative in cosmology is the Big Bang. That the Big Bang was a true historical event is confirmed, both observationally through the microwave background radiation, and theoretically through the *singularity theorems* of Hawking and Penrose which show that all solutions of Einstein's field equations must have a temporal singularity. By a curious combination of circumstances, the Cosmological Principle, a pragmatic simplification of the problems of back-reconstruction, leads inexorably to the biggest catastrophe of all time! The Big Bang is (by definition) the greatest singular event in the history of the cosmos. The catastrophe scenarios of geology are puny in comparison: the asteroid that hit the earth 65 million years ago and killed off the dinosaurs; Brown's hypothesis, that the Earth spontaneously

III. Cosmology

flips its poles every few million years; Velikovsky's "Worlds in Collision," etc. Compared to the Big Bang they are little more than random eye-blinks of an elephant relative to the stampeding of its herd.

Catastrophism combined with the Cosmological Principle may sometimes provide an iron-clad defense to your normally over-defensive cosmologist: *any regularity in the universe is evidence for homogeneity. Any irregularity is a leftover from the Big Bang.*

A few examples of this kind of reasoning cropped up at the Symposium. The autogestion of the galaxies as "islands" in the oceans of interstellar gas was explained by minute fluctuations in the distribution of matter, to the order of 1 part in 10,000. The preponderance of matter over anti-matter was explained in the same way. On the other hand, the uncanny regularity of the distribution of matter in every direction, the so-called *isotropy* of the observed universe, was justified by the homogeneity principle ("homogeneity" implies "isotropy," not the other way around) as was the hypothesis of a unique value for Hubble's constant throughout the length of the cosmos.

Despite my dubious accreditation as a lowly journalist, when the floor was opened after Rees' talk I stood up and expressed the views just presented: namely that a combination of the Big Bang and the Homogeneity Principle made it possible to give an "explanation" for anything. Rees stared at me for a few minutes, then acknowledged that the observed isotropy of matter is "a complete mystery."

I was grateful for the attention paid to my question. Neither my question nor Rees' reply appear in the text of *Some Strangeness in the Proportion*. I am not upset that only the questions of official delegates appear in the transcript. I do think it a bit odd, however, that it would not carry the replies of invited speakers! Over the remainder of the Symposium, Rees continued to give me dirty looks. Once he even banged into me as I was entering and he was leaving the auditorium. Clearly my question had troubled him. The next day he sent around a colleague who, coming directly to the point, asked me what my academic status was. I told him I was a philosopher of science at Columbia University.

Such rudeness was gratifying to my ego, but it was also unnecessary. Serious cosmologists do in fact worry about this unsatisfactory situation. My question was not designed to put Rees on the spot before the assembled world "Nobeliat"! What it reveals is that all of us, professionals like Rees, and yours truly amateur, were grappling with the unsolved difficulties that would lead to Alan Guth's proposal of the inflationary scenario in 1981.

What renders the observed homogeneity even more troublesome is the fact that, by Special Relativity, widely separated parts of the universe are not causally connected. No-one has the slightest idea of what holds the whole picture together. One appears to have a situation (familiar to Quantum Theory which is built on oxymorons) of being obliged to depict the Big Bang as a "homogeneous catastrophe"!

Summary of Cosmology

A summary of the state of cosmology as presented at the Einstein Centennial Symposium:

(1) The microwave background radiation at 3° Kelvin has brought back a kind of fixed reference frame that recalls, yet is quite different from, the electromagnetic ether.

III. Cosmology

(2) General Relativity has survived all falsification challenges by a wide margin.

(3) Irwin Shapiro and John Taylor think that they may have detected the presence of gravity radiation in the energy bursts of binary pulsars.

(4) Dennis Sciama wants us to believe that a Black Hole has been uncovered in Cygnus X-1.

(5) Claudio Teitelboim announced the discovery of a new Observable in Black Holes, the Internal Spin.

(6) The Cosmological Censorship Hypothesis is God's way of fabricating a "cover-up" to the acausality inside a Black Hole.

(7) The observed homogeneity of the distribution of matter in the universe is baffling, and does not in fact simplify the task of observational cosmologists.

(8) Everyone now accepts the standard model of the Big Bang, with essential modifications on the way.

Cosmology is, and always has been, a Paradise of contradiction. Yet one need not agree with Immanuel Kant's "proof" in the *Critique of Pure Reason* that cosmology is a worthless pursuit for serious minds. If nothing else, it keeps bright people off the streets and out of trouble.

Banning the cosmologists from science is akin to Plato's expulsion of the poets from his Republic. It isn't possible to uproot Mankind's unslakable thirst for knowledge about the origins, structure and future of the universe in which He finds Himself. Nor is there any good reason for doing so. He will probably continue to be thoroughly lost, but at least He feels a little better.

Finale

Einstein on the Lunch Menu

Thursday Afternoon: Lunch with Claudio Teitelboim, (theoretician), John Archibald Wheeler (physicist; director of the H-bomb project in 1950); David Malement (philosopher), Adolf Grunbaum (philosopher of science), Mary Wisnovsky (IAS staff, conference coordinator); Martin Klein (historian).

There are parallels between Claudio Teitelboim and Einstein. Teitelboim is also Jewish and a refugee from political oppression. (His father was chairman of the Chilean Communist Party. He survived the Pinochet takeover through the lucky historical accident of attending a meeting in Moscow at the time.) Now he is a fellow at the IAS and had already made several valuable contributions to theoretical physics.

Over lunch Teitelboim spoke out against the amount of time being wasted at the Symposium on "proving" that only Einstein could have invented General Relativity. This gave me an opportunity to expand on my theory of the religious phenomenology at work in this tribal gathering. Teitelboim more or less agreed with me; he'd thought, wrongly it appears, that "intelligent people" would "recognize their emotional tendencies" and compensate for them.

IV. Finale

All of this greatly interested John Archibald Wheeler. He was sitting on the other side of Teitelboim from me, and leaned across the table to hear what I had to say. Wheeler, a very famous physicist, talented expositor of science, and co-author of a standard textbook in the field (*Gravitation*: Wheeler, Thorne and Misner) saw nothing objectionable in myth and ritual. In his view, the organizing committee of the ECS should have set aside a certain number of sessions dedicated to the "myth of Einstein in the scientific and popular mind."

As Mary Wisnovsky was right there, we turned the question over to her. She explained that the human being known as Einstein was being discussed at other Centennial celebrations scheduled around the world, the most notable being the large conference scheduled in Jerusalem right after this one. This was why the ECS was focused on science alone. To my mind this official explanation of IAS policy could be interpreted in 3 ways:

(1) That the "spiritual" Einstein was seeping in through the cracks in any case, as Wheeler, Teitelboim and I were claiming.

(2) The scholars in the "human sciences," habitually snubbed by the IAS right from its establishment in the 30s, were being snubbed once again.

(3) That the only aspect of the Einstein legacy that concerned the IAS was the developments in physics that have provided the basis for the nightmare in which the human race now finds itself. Given the large amount of "overt politics" at the ECS, involving West Germany, Jimmy Carter, Izvestia, Richardson Dilworth, Jürgen Schmüde, etc., it was inevitable that the IAS would want to keep these implications under lock and key.

Mary Wisnovsky met this barrage by dragging out the threadbare blue-ribbon cliché to the effect that Albert Einstein had had nothing to do with the building of the A-bomb apart from his two letters to Roosevelt. Wheeler came to my rescue: he'd made a study of this very issue, and he didn't agree with her. It was his understanding that in the Talmudic tradition (to which Einstein was supposedly heir), the community agrees to support the sage, in exchange for which he bears a heavy responsibility for his acts in the public domain. As spokesperson for the entire community he assumes all of the consequences of his public statements. (It could not escape my notice that these views were being expressed by the director of the H-bomb project at Los Alamos in the 50s!)

Seen in the light of this tradition (of which I am also obviously heir) Einstein believed that, by virtue of a prestige greater than that of any other living scientist, he had played a major role in the launching of the nuclear age. (It didn't occur to me to ask Dr. Wheeler if he thought that FDR was a closet Talmudist!) Mary Wisnovsky graciously retired from the field. She was satisfied that all of these issues would be raised at the Symposium in Jerusalem.

Quantum Gravity; Supergravity

It is a permanent feature of scientific history that great scientists, by virtue of their legitimately acquired reputations, may impede progress in the very areas in which they've made their most notable contributions. The fixed opinions of an otherwise highly original mind may stifle further development of a field up to the day of his death.

Consider the inhibiting effect of Immanuel Kant on the development of non-Euclidean geometries; the rejection of

IV. Finale

Cantor's ideas owing to the opposition of Kronecker; Edison's opposition to alternating current; Eddington's hostility to Black Holes, which almost ruined the career of his most brilliant disciple, Chandrasekhar; the campaigns against vaccination led by some of the leading physicians of Europe.

I am referring, of course, to Einstein's rejection of the Copenhagen interpretation of the Quantum Theory, a theory he'd done so much to develop. In the final decades of his life Einstein's most notable conviction and strongest ambition set up interference patterns with each other, dooming both to failure. The ambition was to create a Unified Field Theory that would unite the 4 known force-fields that bind the universe: Gravity, Electromagnetism, the Strong Force of nuclear binding and the Weak Force of radioactive decay. Had he succeeded, the interpretation of gravity as the Riemannian geometry of space-time could have been extended over all natural phenomena. Every conceivable interaction could then be explained as a wrinkle in space-time. The search for a UFT consumed the two decades of his sojourn at the Institute for Advanced Study.

The conviction, maintained with the same single-minded devotion, was that Quantum Theory was defective. If not totally wrong, then it was incomplete: one could design thought experiments for interactions whose outcomes could not be predicted by its formalism.

Historian Abraham Pais claimed to know the exact month in which Albert Einstein became disillusioned with Quantum Theory: *June, 1926!* His misgivings about the theory had been expressed before, in private letters to Heisenberg and others. it appears that Einstein experienced a panic reaction from reading Max Born's famous communication in which he advocates the interpretation of the modulus of the Schrödinger wave equation as a proba-

Quantum Gravity; Supergravity

bility density. Did the quantum theorists intend to convert the entire universe into a roulette wheel? *God does not play dice with the universe!* With a mournful gesture, Pais pulled out his metaphorical violin and pined:

> "*Einstein's verdict came as a hard blow to Born.... In tears, Ehrenfest ... [experienced] ... a sense of loss, of being abandoned by a venerated leader in battle....*"

Where have all the paradigms gone?

Despite the sterling work that Einstein had done in the area of quantum statistics, despite his having wrested the quantum itself from the confusion of Planck's mathematics, he balked before the final conclusion, a universe statistical in its very essence.

This set up professional barriers to any theoretical physicist eager to combine relativity and quantum theory. No ambitious scientist wanted to risk his academic career by coming up against the unequaled authority of Albert Einstein. Although quantum theory continued to develop, as did general relativity, Einstein extracted himself more and more from the mainstream of theoretical physics, which he himself had created.

Quantum theory did gain a foothold in Special Relativity via a subject called Quantum Field Theory, developed by Dirac and others. Here Einstein's forecast was prophetic: the theory gives correct predictions yet is a philosophical nightmare. No-one had the chutzpah to "quantize" Einstein's field equations in his lifetime. The knowledge that the demigod who had bestowed General Relativity on mankind had withdrawn his seal of approval inhibited all attempts in this area.

This, at any rate, was the story handed down at the Symposium; yet it doesn't sound quite right. For most scientists, the urge to make a name for oneself, to get that

IV. Finale

stiff little photo into the encyclopedias of the future is difficult to suppress. Think of the frequency with which newspapers carry a screaming headline such as: New velocities faster than light. Calculations show Einstein wrong!

What must have really happened is that there were some fierce conflicts between the encyclopedias of the future and the swivel chairs of the present. The swivel chairs won, as they so often do. Quoting from the opening remarks of Yuval Ne'eman:

> "... For sixty years, the *Theory of Gravitation* steered a separate course.... It seemed therefore inconceivable that one should try to undo all that perfection in dealing with its quantum aspects."

This was supported in part by the commentary of Res Jost:

> "Never in the past 50 years has Differential Geometry, the mathematics of General Relativity, entered into as close a link with the theory of elementary interactions, as during the last years...."

The "close link" is in fact the "quantized version" of Einstein's quest for a unified field theory. One way of doing this is to introduce certain structures from Differential Topology known as *fiber bundles*.

I gave a seminar on fiber bundles at the Wesleyan Physics Department last year. They're somewhat tricky and it takes some experience using them to realize that they really are the only way to express the properties of some basic objects in mathematics. Thus, the only correct way to describe a Möbius Strip is in the language of fiber bundles. Once one begins to go into higher dimensional spaces (such as the 4 dimensions of space-time, or the 5 dimensions of Kaluza-Klein theories, invented as extensions

Quantum Gravity; Supergravity

of General Relativity designed to incorporate electromagnetism), the language of fiber bundles, connections, homology, homotopy, gauges, transversality, metrics, derivations, etc., that is to say the stuff of Differential Topology, is indispensable.

And at the ECS adding lots of new dimensions to ordinary space and time was the only game in town! Delegates and invited speakers spoke, with a nonchalant air, of 105 dimensions. Almost all of these, say 101 or so, are hidden in the "fiber." That is to say, the fiber isn't what you make the rug of: the fiber is what you sweep under the rug! Turning an apt metaphor into a deliberate catachresis, if Black Holes are described as having "no hair," then Supergravity is a "superhair" theory! But what are all those extra dimensions doing there?

My understanding of what Yuval Ne'eman was telling us is they aren't really there. In line with the physics instructor from SUNY who maintained that fields are nothing more than convenient calculating devices, the dimensions in the fiber are only introduced to give the equations some much-needed symmetries. The method is familiar to mathematicians in the form of the Lagrangian multipliers developed in the 18th century. It has received new life through the important theorems of the great mathematician Emma Noether (another German refugee to end up in a Philadelphia suburb, Bryn Mawr. Though neighbors for a short period until her tragic death on April 14, 1935, she and Einstein never met). These theorems state that every symmetry principle in the differential equations of physics can be interpreted as a conservation law in the universe.

Like catalysts in a chemical reaction the extra dimensions can be taken out again after the calculations have been made. In fact Ne'eman had a term for them: he called them *"the ghosts."* Not everyone thinks that the ghosts aren't real. The way Tullio Regge describes it there are 6 ex-

IV. Finale

tra dimensions which we have to consider as components of the real world: "Why pretend any longer?" he pleaded: "We really live in a 10-dimensional universe!" Into these 6 extra dimensions Regge has squeezed the four forces of nature and some essential tensor fields.

Ne'eman's talk boomed with great mathematical cannons. If Plato was really serious about excluding persons ignorant of geometry from his Academy, here is a sample of how hard it might be to pass the entrance exams these days:

$$\delta\chi^a + \chi^{at}\chi^t\Lambda + \frac{1}{2}\chi^c\chi^d R_{cd}{}^a\Lambda$$
$$+\frac{1}{2}\chi^{cd}\chi^{ef}R_{cdef}{}^a\Lambda$$
$$+\chi^c\chi^{de}R_{cde}{}^a\Lambda = 0$$

$$\delta\chi^{ab} + \chi^{at}\chi^{tb}\Lambda + \frac{1}{2}\chi^c\chi^d R_{cd}{}^{ab}\Lambda$$
$$+\chi^{cd}\chi^{ef}R_{cdef}{}^{ab}\Lambda$$
$$+\chi^c\chi^{de}R_{cde}{}^{ab}\Lambda = 0$$

Concerned that these equations might be confusing to some of us, Dr. Ne'eman brushed them aside and flashed another transparency onto the screen:

Dali's Dripping Watches

This, he proclaimed, is the *softening of the fibers!!*

His talk was followed by Peter van Nieuwenhuizen's. His colorful account of the "fibers" depicted them like stiff toothbrush bristles or strings of cooked spaghetti. They can be wrapped, twisted, shrunk to a point, anything but sliced. Through all these transformations, abstract principles known as "super-symmetries" are preserved. To get the elementary particles out of them one has to bring them down to earth by breaking them; hence the term "broken symmetries." This procedure elicits all the familiar elementary particles, along with some new ones, such as gravitons. No-one's ever seen them, but Niewenhuizen charged ahead anyway, with *gravitinos!*

Relativity's Future, in a manner of speaking

As a kind of re-run of the effect of Newton's discoveries in the 17th and 18th centuries, Relativity is once again forcing physicists to become higher mathematicians. In certain regards they've become more "qualified" than the mathematicians themselves who, for unrelated historical reasons, have gone along the road of extreme specialization. After the labor of mastering one or two specialties most mathematicians balk at the prospect of diversifying into other fields. Yet that is what an investigator in general relativity must do today. He must have a command, or at last a working knowledge of: Projective Geometry, non-Euclidean Geometry, Differential Geometry, Differential Topology, Operator Algebras, Partial Differential Equations, Lie Algebras, Probability and Statistics... the list goes on and on.... Where it leaves off is where he must begin tackling several branches of physics!

The relativist of the future will begin his education at the cradle. At age 20 he may be permitted to ask a few questions about the relevance of all that he's been learning. His

IV. Finale

education completed at age 50, he makes a minute contribution to science in his 60s, then dies at 65—from exhaustion!

I Become a Filmmaker

Thursday Evening: The delegations are leaving. Some are flying to Jerusalem. I've joined the team of film-makers. The problem of returning to upstate New York has been solved. From Sunday through Thursday I wandered about the precincts of the ECS with 20¢ in my pockets. As a professional consultant to the film crew I now collect $30. Marty Fuller, Buckminster Fuller's nephew, is the director. Invoking the Golden Rule of the freelancer—*never turn down a free meal!*—we hang around for the final dinner. Everyone at our table is working for the media. Paul Dirac and R. Dicke (inventor of a theory of gravitation with important differences from Einstein's) are seated at the table directly opposite ours. Off to the right sit Rees and Sciama, huddled in conference.

Salade Macedoine, followed by Tournedos, with small baked potatoes; two kinds of wine. We talk and talk: France in 1968; Buckminster Fuller: "Doing more with less"; The University of Pennsylvania both of us regard as the ultimate traumatic experience; the stupid footage the team is required to film. I explain the concept of "zero-point energy" to them: one of those useful notions from Quantum Theory that can be applied to daily life. A young French cameraman shares his philosophy of life: *I chain smoke and think about death.*

It is like an afternoon on the beach after a prolonged winter. Around us the scientists continue with their monotonous stew of polite conversation and endless shop

I Become a Filmmaker

talk. Thus driven to nonconformity, our table is extroverted, rude, boisterous, vulgar. I begin to understand why film personnel make every effort to appear asinine in public. There is a particular form of alienation involved in forever standing on the sidelines while making films about others. To adapt a concept from Yuval Ne'eman talk: we reside in the tangent plane.

IV. Finale

Homecoming

Friday, March 19, 10 AM: I am back in Dutchess County, taking breakfast in a small restaurant near the Poughkeepsie train station. The Loop Bus that will take me the final 30 miles to my destination leaves in less than an hour. My briefcase, fat with press releases, reprints, lecture notes and other materials, has been placed under the table out of sight. For the first time in a week the surrounding clientele consist mainly of construction workers and small-town businessmen. Already I'm wondering if there is anything in my report that might be meaningful to them.

For the next two days I will be enjoying the mere sensation of being alive. Thereupon begins the labor of sifting through the precious booty from this latest of my many expeditions, my "voyage-projects." After they've been read, the reprints will go to Peter Skiff, the physics teacher who tried to discourage me from attending the ECS. The 3 notebooks assembled on my desk will be studied, assayed, carefully examined from every point of view. I will be looking, not so much for scientific insight (although this too will be important), but for human responses to human situations. Connections will be sought through history, times, places, people, amplifying the resonance of broad historic themes.

More than anything else I will be looking for Einstein. In different times and places at the Symposium I caught a glimpse of the real man—like a flash of quanta escap-

V. Homecoming

ing from the accumulation of speeches, at the heart of an ingenious hypothesis, in an after-thought from some ponderous or self-serving political tribute.

At times one fancied him standing in the shadows, though all eyes were fixed on the stage. Facets of the legend that never appeared at the Symposium appeared on the road there, and the way back.

Many were the Isaac Rabis who brought with them into its laboratory the Einstein they'd come to find. I hope, and still hope, that I was not one of them.

Spring 1979
Spring 1987
Spring 2005

Bibliography

[1] *Some Strangeness in the Proportion: A Centennial Symposium to Celebrate the Achievements of Albert Einstein*; Edited by Harry Woolf; Addison-Wesley Publishing Company, 1980

[2] *Albert Einstein, Historical and Cultural Perspectives. The Centennial Symposium in Jerusalem*; Edited by Gerald Holton and Yehuda Elkana; Princeton University Press, 1982

[3] *The Structure of Scientific Revolutions*; Thomas Kuhn; International Encyclopedia of Unified Science; U Chicago Press; Vol 11, #2; 1970

[4] *The Principle of Relativity; A Collection of Original Memoirs on the Special and General Theory of Relativity*; Dover Publications, 1952

[5] *Subtle Is The Lord... The Science and Life of Albert Einstein*; Abraham Pais; Clarendon Press; 1982

[6] *The Large-Scale Structure of Space-Time*; S.W. Hawking and G. F.R. Ellis; Cambridge U.P., 1973

[7] *Gravitation*; Misner, Thorne and Wheeler; Freeman, 1973

[8] *Who Got Einstein's Office? Eccentricity and Genius at the Institute for Advanced Study*; Ed Regis, Addison-Wesley, 1987

[9] *The Transit of Venus: A Study of 18th Century Science*; Harry Woolf; Princeton UP, 1959

[10] *The Einstein File*; Fred Jerome; St. Martin's Press, 2002

[11] *Einstein in Love*; Dennis Overbye; Viking, 2000

[12] *Principles of Physical Cosmology*; P.J.E. Peebles, Princeton UP, 1993

About the Author

Roy Lisker was born in 1938. In 1954 he entered the University of Pennsylvania school to work on an advanced degree in mathematics. Within two years he discovered that the call of arts and letters was stronger. He began working at fiction and non-fiction in 1958, returning to the University of Pennsylvania in 1962.

He has been published in the United States, France, England, Canada and Ireland. He has worked in fiction, science writing, math and physics research, criticism and journalism. Since 1980, he has been the author and editor of several privately subscribed newsletters, culminating in 1985 with *Ferment*, which existed for twenty years as a paper publication before going on-line as *Ferment Magazine*.

In Memoriam Einstein, written in 1979 and first published in French in 1980 by *Les Temps Modernes*, the magazine of Jean-Paul Sartre and Simone de Beauvoir, was the first of a series of Voyage-Projects. These spontaneous journalistic adventures, undertaken usually with very little financial backing, formed the substance of much of the material for *Ferment*.